With love to Brooke, Burley, & Erwin.

The
Baseless Fabric
of this
Vision

A Quantum-Field-Theoretic
Model of Consciousness

Casey Mitchell

Sophia's Ichor

Copyright © 2022 by Casey Mitchell

Published by Sophia's Ichor
sophiasichor.com

All rights reserved.
ISBN: 978-1-7781126-0-7

The Baseless Fabric of this Vision

"Our revels now are ended. These our actors,
As I foretold you, were all spirits, and
Are melted into air, into thin air:
And like **the baseless fabric of this vision**,
The cloud-capp'd tow'rs, the gorgeous palaces,
The solemn temples, the great globe itself,
Yea, all which it inherit, shall dissolve,
And, like this insubstantial pageant faded,
Leave not a rack behind. **We are such stuff
As dreams are made on**; and our little life
Is rounded with a sleep."

— William Shakespeare, *The Tempest*,
Act IV, Scene I

Contents

Introduction — xiii

Part I - The Flame of Cerebral Commotion

CHAPTER I - By the Light of Awareness — 1
CHAPTER II - Spheres of Being | Fields of Sense — 8
CHAPTER III - On Matter | Being Known — 26
CHAPTER IV - On Mind | Knowing Being as Thinking Matter — 33
CHAPTER V - *What it's Like* | A Primer on the Phenomenontology of Consciousness — 42

Part II - The Baseless Fabric of this Vision

CHAPTER VI – Immateriality at the Core of Quantum Theory — 59
CHAPTER VII - Mythologies of the Quantum Foam — 81
CHAPTER VIII - Maxwell's Angel – Aether: God of Light — 87
CHAPTER IX - Invariant Angel | Malleable Spacetime — 98
CHAPTER X - The Quintessence of Quanta — 104
CHAPTER XI - Quantum Waves | A Torus of Tides — 121
CHAPTER XII - Summing to One - I | Superposition — 133
CHAPTER XIII - Summing to One - II | Entanglement — 140
CHAPTER XIV - Chimes of Kaleidoscopic Color | The Music of the Spheres — 145
CHAPTER XV - Indeterminacy | Heisenberg v. Schrödinger — 153
CHAPTER XVI - The Strangeness of Spin | Ouroboros & The Seed of Self-Reference — 157
CHAPTER XVII – How to Kill Zombie Cats — 176

Part III - The Mindfield

CHAPTER XVIII - The Field Whose Property is Awareness — 197
CHAPTER XIX - *Creatio Ex Nihilo* | Efflorescent Complexity and the (Bi)Unitary Nature of Information — 208
CHAPTER XX - Holography | To Write the Whole — 213

Glossary — 230
References & Bibliography — 241

"If the One is not, then nothing is." - Parmenides

Introduction

Mythical thinking, as a precursor to empirically verifiable, rational thought, hypostatizes a "primordial substance" to be comprehended as the void-womb of the world. This pre-Big Bang referent goes by many names; the ancient Greeks called her Khaos. A "chasm of air" said to "gape wide open," Khaos is the primal principle of allowance, a "That" which let's fountain forth all Being as infinite Becoming. This 'indeterminate nothingness,' thought to pre-exist even existence itself, is Darkness Absolute; an abyss without bottom.

Later, in the 6th-century B.C., the pre-Socratic philosopher Anaximander conceptually transforms this non-entity (Khaos) into the *apeiron,* a word whose etymological roots mean "without end." Boundless and indefinite, Anaximander tells us the apeiron is a "thing" that is no thing; an indeterminate, massless mass, and generator and retainer of all opposites. With no differentiable "parts," this timeless, elemental substance, lacks all constituting distinctions such that even causality, space, and time have no meaning with regards it.

When we look back on these primitive ideas, we see them as archaic. But seeing them so is disingenuous to their importance, as it is contemplations like these that cause us re-cognize and reconsider the generative origins of this mystery we live. As time would have it, this idea of a "primordial substance" is still with us, only now we actually know a thing or two about it. Today, we know the apeiron by another term: **quantum foam**. Despite the change in name, this doesn't mean we know *what it is*, at least not in any way that would make sense to rational thought.

To top it off, quantum mechanical theory will not, in fact *cannot,* tell us "What" it truly is, but it does hint at a few of its characteristics. That is, we know some of its attributes as ubiquitous **quantum fields**. Timeless, ubiquitous fields of effect that permeate the quantum foam and serves to point towards an understanding of the vacuum of space as an ocean of seething potentiality, a kind of determinable but as-of-yet indeterminate nothingness. A 'something' not unlike the ancient Greeks concept of Khaos-cum-Apeiron.

Beyond rational thought, a transcendental mode of human consciousness can know this timeless Substance *as* Sublime.

This is a book that seeks to answer foundational questions and is therefore about the *foundations* of physics and philosophy of mind. Its motivation arose out of a rekindled wonder about the natural world. This renewed interest was generated by a surprise, as I, in my naiveté, once believed I understood all there was to understand about the world; such is the hubris of youth. That was until, in a moment of deep depression, I witnessed with unshakeable clarity, a rare and novel kind of knowing. I was surprised by my utter unfamiliarity with such a state while also feeling entirely liberated by it.

In an altered state of consciousness – the one that lies at the root of *cosmic-religious-feeling*[*] – I witnessed the Unity of Reality as an Epiphany into the Nature of the Holy. In the East, this modality of mind (or altered state of consciousness) goes by the names: Nirvana, Samadhi, or Brahman. It is the Prajna-Paramita of the Buddhists, the sacred aim of their meditative practice culminating in That which is beyond-all-knowledge; the point at which subject and object are no more. While in the West, following the philosophical school of phenomenology – which studies the invariant, architectonic structures of consciousness – it is referred to as a vertical "mode of givenness."

In this state of mind, the experiencer and what-is-experienced become One. The normally "two" poles of total consciousness – subject/object, self/other, Home/Alien – unify into a single felt impression and one comes to "identify" their "self" with the *totality of consciousness* rather than merely relegating itself to its most familiar pole. That this kind, type, or organ of knowing is rare causes it to suffer from being all too easily dismissed. This is unfortunate as it grants to its privileged witness a novel kind of knowledge that is unattainable any other way. That is, an intimate knowing as regards the absolute Unity of reality. But reality is not "given" to us finite creatures as a unity, but instead as made up of indescribably many objects. The trouble then becomes explaining how it is that reality is, in some deeper sense, One. We have here the problem first articulated by the pre-Socratic philosophers concerning the Many and the One.

It is my contention that modern science provides the answer: it is to be found in the structure of *how quanta relate to their fields*. Not only this, but as we shall see, **quantum field theory** supplies us with an adequate *model* of what consciousness may be: a timeless, omnipresent, massless *field*. A characteristic of the vacuum that is accessed and actualized by sufficiently complex aggregates of energy. In a word, by living organisms – the *élan vital* of Life being the property of space (or "force") that expresses the field.

But before we arrive at a quantum-field-theoretical model of consciousness many preliminary remarks must be made. I must explain to you, dear reader, my philosophical commitments. That is, supply an ontology concerning what I

---

[*] This is Albert Einstein's term for this rare state.

understand to be indisputably real. That I take consciousness, mind, and its psychological "objects" to be real — as real as the atoms studied by physics, say — does not mean I wish to espouse an idealistic metaphysic. The physical world that exists apart from the observer goes on existing without its being observed. With regards the ontology I will set forth, both mind and matter have an equal place within it as they are both borne of *fields*.

For ease of argument, we begin with assuming the dualism of Descartes: both mind and matter exist but *prima facie* appear to possess irreconcilable differences. 'Cognitive' substance, — his *rez cogitans* — is immaterial, ethereal, and formless while 'extended' (or better, material) substance — his *rez extenza* — is just that, extended in space, material, tangible, and structured. The question is: what must reality be "made of" such that it can entertain with equal existential weight and value, objects that belong to, and are generated by, each "substance"? Both deserve equal treatment.

To reflect upon the nature of mind, we must invert its eye, in a strange-looping, self-referential way, such that it shines upon itself the light of its own illuminative knowing. Is that even possible? Indeed, the culminative achievement of the "turning of the light of consciousness upon itself toward its source"˙ results in an enantiodromedal transformation wherein the subject of experience dissolves into its objects such that "consciousness" becomes All that Is. In other words, the observer and the observed become One.

Of matter, primarily, we *know* it via the mind as a "restrictive" *substance*. Because the material furniture of reality is public and thus the same for everyone, we call it "objective." The web of its self-relations designates a reality we need not *necessarily* be a part of — the sun, moon, and stars, relate as they do *with or without spectators*. At the same time, what we commonly call matter is collectively conceived by many conscious agents as a unity, and in this way, it forms the material-energetic system we call universe, the Soul of which — *energeia* — can never be destroyed. It may be trite to say, but it warrants expressing; no one, not any kind of observer whatsoever, in the history of the universe, has ever known anything "outside" of their consciousness. By this self-evident admission and in this very sense, mind can be seen as both primary and fundamental. An insight we might codify with the aphorism that there exists *mind under matter.*

Regardless, to comprehend matter as it is in-itself, we must turn to our best physical theory — quantum field theory (QFT). QFT is a deep theory, the unification of classical field theory, quantum mechanics, and special relativity. Although the very question QFT is thought to address — that of the true nature of matter — we shall see it offers no clear commitment to any physical **ontology** whatsoever. Instead, every one of its many "interpretations" — each possessing

---

˙ Merrell-Wolff, F.: *Transformations in Consciousness.* (1995)

its own ontological and epistemic flare — hinge on answering *just what matter is*. Familiar readers might claim that it is the proposed resolutions to the famous **measurement problem** that spawn the theories many interpretations, and they are correct. The measurement problem is best understood in light of the **double-slit experiment** for it is here that we most clearly see the problem of "**wave/particle duality**" that haunts the theory. You see, experiments only ever result in single, particle-like, impact points, everyone of which a definite outcome. This hints to us that the quantum is a particle. But theory predicts indefinitely many outcomes and is itself based on waves and it is the statistical probability of wavelike phenomena that emerges after many accumulative trials. After analyzing a conglomeration of these impact points, we observe the tell-tale signature (an "interference" pattern) that hints to us that the quantum is a wave. This duality is a deep ontological problem.

As such, the measurement problem is more deeply the issue of trying to ascertain just *what* the fundamental constituents of the theory *are*. To use Bell's term, what are its postulated 'beables;'˙ *waves* in and of *fields*, dimensionless point *particles* floating against a backdrop of truly empty space, some exotic *wavicle,* or are they something entirely different? As we'll see, we must take from both concepts the relevant aspects that experiments appear to show as actual such that we arrive at the novel concept of the aggregable **quantum**; the unitary entity that transcends both its wave and particle forebear's and behaves in seemingly impossible ways.

Again, the formalism makes no statement as to what its actual beables are although it does give us three clues as to what they may be; one, they possess an aspect of discrete individuality, two, can only sum as aggregates through superposition (this will be explained later), and three, they lose their individuality and become indistinguishable from one another when they entangle to form larger systems... even though the larger-order system itself retains its 'beable' nature, that is, it simply becomes a larger expression of that which makes it up. It is these three clues that will lead us to the most sensical interpretation and model of what the quantum actually is, this will in turn tell us what matter actually is.

But what of those interpretations? **Pilot-Wave** theorists assume the simultaneous existence of both waves and particles. They interpret the concurrently present characteristics of these incompatible beables as a "pilot-wave" that guides a particle to its definite outcome.

The **Copenhagen interpretation** confines us to language-generated intersubjectivity and, as such, is an **instrumentalist** interpretation. This construal says something entirely different: that we cannot possibly know what

---

˙ This is J. S. Bell's open-ended term for the theory's fundamental constituents as we know quanta are neither wholly particles nor waves but something more exotic.

reality even is, that the best we can do is calculate the likelihood that a certain event will occur and all that we know is the theory as accessed by rational thought. This interpretation does not get us to the way in which the world really is but instead leaves us trapped in our minds and its epistemic structures.

The **Many-Worlds** interpretation of tells you nothing about "what" the world is *made of* but rather astonishingly claims that every conceivable — even every inconceivable — reality exists. Stated differently, any possible event (or definite experimental outcome), given an initial state, *can and does actually happen*, although each result relegates itself to a bifurcating "branch" of the ultimate **wavefunction**. Unlike the Copenhagen interpretation that speaks of our access to the world, many-worlds is a theory about reality as a whole.

Now, just what is this "wavefunction?" Normally, standard procedure treats the wavefunction as a "mathematical object" meant to overlap with and informationally describe the quantum. It may or may not be informationally complete and either describes a quantum system wholly or partially. Realistic interpretations of this function treat it as **ontic** (the wavefunction veridically represents the quantum state) while instrumentalist interpretations treat it as **epistemic** (it represents only our possible knowledge *about* the quantum state).

The **Bayesian interpretation**, or **Qbism**, like the Copenhagen, commits itself to the notion that we can *only* examine the inside of the veil, as it were. That there exists a 'transparent cage' about us such that we are confined, knowingly, to the actions of conscious agents and their epistemic pastures, to the "subject" side of the subject/object relation. The 'observables' of quantum theory becoming just that, things that are only ever observed *by observers.*

None of these, I believe, are satisfactory.

However, lying "underneath" these hides the quantum **information-theoretic** interpretation, where we arrive at the "it" of the theory — its beables — "from bits" of information. This interpretation is almost-certainly, at least partially, correct whose only fault is that it doesn't go far enough. In fact, every quantum interaction represents the way in which the universe computes itself, indicating the "flipping" of its bits as a local exchange of information thus keeping track of its increasingly complex evolution. This reading further accentuates the immaterial aspect of the theory, as well as validating its many discontinuous features, but again, suffers from a lack of commitment to a "substantial" beable. That is to say, do the bits produce waves or particles? We will examine this "machine-code computational processor layer" of the universe in the later chapters of the book.

Now, of these many interpretations there are only a few that treat the theory realistically. That's what we want, to know and model that portion of reality that — although constitutes our being here — would nevertheless obtain *without us.* Achieving this requires subscribing to some form of wavefunction realism. Many-Worlds does this but its ontology of infinitely many copies of reality only serves to undermine its value and so must be set aside.

Objective Collapse (OC) is the interpretation that I will be arguing for as it maintains that, put simply, the wavefunction is real; that the quantum state *obtains*, that is, acquiesces to a physical, observer-independent reality. As a model, OC forms the greatest explanatory apparatus, and resolves many of the apparent paradoxes by appealing to a *realistic* interpretation via a fundamental *fields* ontology. The only difficulty with this interpretation is accepting the somewhat *ad hoc* procedural dynamic know simply as **collapse**. But this is not without merit, for collapse is what we observe in experiments. As a quantum characteristic, collapse is best understood when couched with its meaningful referent, as either "quantum state" or "wavefunction" *collapse*. Here, it is recognized as a global, discontinuous *change* of state, where the wavefunction "localizes" itself by undergoing an immanent and immediate restructuring.

Some other merits of the OC interpretation is that it allows us to coherently model the Many and the One; to demonstrate how there can exist a multiplicity of "unconnected" individuals or spatially separate objects that gather to from a whole. It can also substantiate the relationship between the continuous and the discrete. For you see, OC paints reality as a unified plenum, an ocean of timeless property spaces (fields) that are capable of entertaining discrete, albeit nonlocally-extended, energetic "units" of themselves. We have met these items; they are called **quanta**. The second part of this book is dedicated to establishing this model.

To properly appreciate OC, I have mentioned the importance of stressing a fields ontology, but just what exactly is a **field**? Fields are most easily intuited as immaterial *properties of space* that manifest as physical effects. The mathematical definition of which is a function over space and time. Einstein's **general relativity** is often understood geometrically, as the "curving" of space and "warping" of time, but it can equally be framed as a field theory wherein the dynamics of spacetime itself form a field of gravitation. Here, "gravity" is the fields property. In this view, it is known as the *metric field*. The word 'metric' comes from the fact that this field has geometrical symmetries and invariants that one can use to define distance and is better modelled as a kind of *fluid*.[*]

It may be beneficial to view all fields as various kinds of fluids, for they serve as a medium for waves and other distortions to propagate through them and bilaterally warp the structure of the objects they contain. However, despite their fluid-like *qualities*, we must remember that fields and their "ultimate carrier" are "made" of immaterial waters.

Again, physically, the world as we find it presents itself to us as a veritable phantasmagoria of phenomena. A phantasmagoria that presents itself as a philosophical puzzle — The Many and The One.

---

[*] Wilczek, F.: *A Beautiful Question*. Penguin Books. (2016)

QFT's may hold the conceptual keys because of their structure. A field is a unified and continuous whole that entertains discrete resonant motions of itself, each one of which a coherently stable, spatially extended pattern — a wave. The field is an ocean and the waves it enables are unified and individuatable quanta of rippling energy. All this is to say that the oceanic ontology of QFT allows for the One to remain Whole while being able to accommodate parts. With respect to a fundamental ontology, Sunny Auyang has rightly pointed out that *only* fields allow us to coherently picture a whole differentiated into parts. She writes *"A field is a genuine whole comprising genuine individuals, a continuous world with discrete and concrete entities..."*[*]

Both fields and their waves are continuous entities except a quantum as unified wave possesses a discrete energy. The discreteness of which is what grants to its nonlocal extension a "unity that overleaps space."[†] It is this discretization that allows the field to entertain "separate" individuals, i.e., *different* quanta. "Separate" is here in scare quotes because it signifies an illusion, for in reality, all quanta are expressions of one and the same thing — the field. Therefore, all field quanta, all of its "individual objects," are really One and the same underneath.

What does all this have to do with mind and matter? Well, in the end, the question will become, how do mind and matter *relate*; do their intrinsic nature's commute? That is, do they compliment one another, or do they remain fundamentally and irreconcilably distinct? I will show that an OC interpretation of QFT with corresponding commitment to a field/information-theoretic ontology can and does place both mind and matter on equal grounds. From here, both can be seen to share in the self-same intrinsic (in!)substantiality. Mind, in this picture, need not only be seen as a kind of quantum field but also as an integrated-information[‡] processor.

Given all this, the broad scope and aim of this book is to merge what we commonly think of as "ideal" with what we commonly consider real — to place mind and matter on equal footing.

In the first part of the book, we shall philosophically address reality-as-it-is-lived-and-known. Here, "consciousness," that "flame of cerebral commotion,"[§] will be our underlying theme as I will try to place in view what "reality" is, what's "real," and what counts as existing.

---

[*] Auyang, S.: *How is Quantum Field Theory Possible?* Oxford University Press. (1995)
[†] Eddington, A.: *The Nature of the Physical World.* The University Press. (1928)
[‡] Integrated Information is a theory of consciousness put forward by Giulio Tononi.
[§] This poetic description is taken from Will Durant.

Following this, in the second part of the book, we will move into attempting to understanding the external world, or, as I like to utter with a realistic bent, *that*-which-exists-apart-from-us. Here, rather curiously, we will be considering our best physical *theory*. A curious situation as it gives credence to instrumentalist approaches to quantum theory and the notion that we only ever know our mind and its contents. In this "curious" way, it is always and only a *theory* of *that* which exists apart from us. Regardless, in this part, we will ignore the philosophical straitjacket that argues for the primacy of mind and instead become pre-critical metaphysicians and focus the aperture of our understating onto the nature of matter as naively understood. The theme of part II is to look to QFT to examine the immateriality of reality and establish what is meant by the book's title.

In the third part, we will put forward a theory of consciousness based on all that we have learned of the physical structure of reality. We will apply these principles to the mind and see how it resolves much in the philosophy of mind. We will also investigate the holographic nature of the universe and consider the special form of consciousness that supplied the motivation for this project.

Finally, please forgive me for the somewhat jumpy thread of the main text. This book has taken the better part of 6 years to put together and much of it was written in a time when I was ignorant of many things (and still am). Also, given that this is a self-published book, it lacks the professionalism of a copyeditor. Nevertheless, I have tried my best to keep it as concise, accurate, and as physically relevant as possible, but there are most certainly interpretational mistakes and psychological prejudices that will have found their way onto the page. I kindly ask that should you find any grammatical or punctuation errors that you email me at caseywmitchell@yahoo.ca so that I may fix them. Or even if you have a question or would like to send me a comment. Lastly, I thank you for your interest and hope you enjoy the book.

A final note, there is a glossary at the back for bolded words.

Part I

The Flame of Cerebral Commotion

> "What greater thing is there than this Mystery
> that is Myself?" — Franklin Merrell-Wolff

CHAPTER I

By the Light of Awareness

Do you ever wonder what reality is? Or wonder why, even after all this time, existence remains a mystery? Just what is the true nature of Nature and how does mind fit into this seemingly "natural" order? If matter and light make up everything we take to be indisputably *real*, where does that leave that flame of cerebral commotion we call "consciousness?" That candle that *illuminates* and *knows*.

Without consciousness, there would be no wonder, no social reality, no knowing, and no Love. From what is its origin and how does it relate to living organisms? How are we to take seriously the modern neuroscientific understanding that 'consciousness' is somehow 'equivalent to' or 'generated by' an intricate web of firing neurons? That the physical pangs we suffer and the sorrow of heartache we carry is somehow 'nothing-but' an electro-chemical soup formed by our brains and nervous systems? Or could it be the other way around? Could the true nature of reality be in some way related to the mind? Could the phenomenon we are here calling "consciousness," have been present at the dawn of time and part of, if not wholly, the universe's very fabric? This book claims just that.

To say that we find ourselves in a complicated, multifaceted reality might be the ultimate understatement. Not only is the world apart from us chaotic and variated, but even the cognitive faculties and sensual modes of knowing that gets us in touch with it are themselves varied. Indeed, human beings are exposed to many equally actual, truly existing, domains of experience. Empirical experience, pre-reflective awareness, or simply sense perception, perceives reality-as-it-is[*] with many sensual modalities (visually, auditorily, etc.). This is knowing by living through. It is a form of experience that is a

---

[*] Not quite. Reality given through the senses has been distorted by its informal processing. However, the representation thus revealed is still part of actual reality. By this I mean our naïve mental states, although partial, still obtain as actual.

prerequisite for, but not guarantor of, achieving access to the conceptual domain of rational reflection. A domain capable of revealing more profound truths about the world.

Naively understood, perception gets us in touch with the "real" world and its many unprecedented material forms. Although what we perceive directly is our brains "best-guess hallucination," that "hallucination" itself is real and it reveals a structure that is certainly already there, an environment ready to be perceived. That is, despite some experimental phenomena revealing the structure of the hallucination – the saccadic movement of our eyes and the brain's constant but unconscious filling in of a blind-spot – the brain, from its own dark recesses, "projects" an image of objects and entities that are confirmably "there." As such, it will be argued that we perceive the world naively or directly-as-it-is. Those prior examples being minor "patches" to otherwise accurate visual perception.

We are not limited to sense perception, but as language-endowed, rational agents, we are also privy to thoughtful conception, and through it we gain access to the Infinite. As a quick note, in what follows I am not critical with the use of the word 'experience.' For me, it denotes simply 'being aware of' or 'living through,' and as such, I use it as a synonym for consciousness. Historically, the term 'experience' was reserved for perception as sensually perceived actuality; first-order knowledge given through, or arrived at, by way of the "physical" senses, such as seeing that something is the case.

The philosophical school of empiricism restricts experience to this empirical sense and may now be contrasted with rationalism. Rationalism denotes the use of employing reason as a method of arriving at knowledge – second-order knowledge. Here, one exercises the faculty of thought to arrive at novel understanding. We don't experience thought as we do the local environment. However, this distinction is irrelevant where consciousness is concerned. Regarding it, anything that may be illuminated by it is at the same time experienced by it. Because it stands behind both, consciousness has "equal access" to both sense perception and rational cognition.

The capacity to conceive, to rationally reflect, is a profound, albeit fallible faculty that enables us to understand the world in ways no other creature can. To 'perceive' means to "thoroughly grasp" and receive the given as it is given. To 'conceive' means "to take into the mind and form a correct notion of" what is beyond the given. Although perception and conception encapsulate much of what the human mind can experience and understand, it is not exhausted by them. Other *modes of knowing* are available to our complex form of consciousness. Structures that exceed these two modalities' bounds such as self-consciousness, "altered" or "mystical" states, dreaming and its lucid phase, and the inner bodily senses of tachtio and proprioception. Furthermore, we can reflect on reflection, we can think about thinking.

Somewhat worryingly, all of these modes are bound to their own spheres of evidence and can fall prey to error and illusion. We can misperceive and falsely believe, but we should not fear this fact as it is a silent testament to our freedom and serves as a condition of possibility for our getting it right. To live in err is a privilege because it is granted to us by our free agency.

Perception (pre-reflective, empirical experience) and conception (rational reflection) form the most basic modes of our knowing with regards consciousness. But a more cogent and clear distinction as to what is revealed by them is gifted to us by the philosopher Wilfrid Sellars when he speaks of the 'manifest' and 'scientific' images. Each encapsulates a different way we know the world.

By 'manifest,' he means what is immanently available to us as human-level, conscious creatures; what we experience directly. It is the world that dawns upon and happens *to you* and as such *includes* social objects as their meaningful actuality barrels into our minds on the wings of words. For instance, the reality and existence of morally-bankrupt company Apple is seen for what it is when it is mentioned to us — that is, made manifest — by others. The manifest image of Sellars includes anything from tables and chairs to plants and animals, companies, and corporations to promises and platitudes.

By 'scientific,' he has in mind the world as understood through the lens of modern science. An image wholly constructed *by us* meant to overlap with the preferred object of our study — Nature. Here, we know the 'what' of the world through our experimentally verified, mathematically backed *theories*. In this image, we have uncovered the fact that nature employs **quanta** — energy-momentum-charge densities — that flow through "operator-valued" quantized *fields*. These are complex, nonlocal waves that, through their dynamic interactions become localised oscillatory movements — atoms and molecules. These in turn, structure the furniture of reality, and when converted to radiation, ignite it in chromatic light.

As a precursor but not to go to far off-track; intrinsic to the very depths of reality, *fields* are fundamental. As properties of the **vacuum,** they permeate and interpenetrate the whole universe. Each particular one of which harbors an elementary quantum, a wavelike, individuatable unit that is either **resonant** or **virtual**. A resonant quantum is orderly and rings out as a fields natural tone while a virtual one only chaotically disturbs it. A virtual quantum is more like a field's unnatural disturbance.

There exist only two types of fields with corresponding quanta: *matter fields* populated by **fermions** and *interaction fields* populated by **bosons**. Due to certain symmetries and conservation laws, the interaction fields exist *necessarily* and are more commonly called "force" fields. Interaction fields facilitate the interactions of matter fields, and their contribution appears as a "force." Only two kinds of fermions make up all the matter in the universe —

quarks and electrons, while gluon and electroweak fields supply an arena for them to interact without ever touching. It is reasonable to consider every physical object as an amalgamation of quanta resonating certain fields in various ways. Here, matter is understood as a portion of the field that is vibrating in such a complicated way that its characteristic actuality emerges as solidity. Said differently, matter is a state of the vacuum that has more energetic vibrations at a particular location. Matter is nothingness set into motion.

The Modern Stance of Reflection

Both images (the scientific and manifest) — in fact, anything of which we are capable of grasping or coming to know — will only-ever and always be thrown up and illuminated by that flame of cerebral commotion we call "consciousness." A transparent reflection, consciousness is the clear mirror of the world, and it is *only* by its light that anything can be *known, sensed, or felt,* whatsoever.

This does not mean that consciousness props-up or fabricates the whole of *existence* as that actuality is a prerequisite for its coming into being. Further analysis tells us that in considering consciousness that we are dealing with an indefinable somewhat, a thing that is no "thing." Yet it is a "thing" of such a foundational nature that its actuality is assumed in, not only any attempt to define it, but in every human affair whatsoever, including the writing and reading of this very sentence.

"If we take eternity to mean not endless temporal duration but timelessness," Wittgenstein tells us, "Then eternal life belongs to those who live in the present." Consciousness exists in that ever-rolling window we call the present and it establishes itself in the sliver that cleaves self and other, on the periphery that defines the difference between a sentient agent and its environment.

An "intentional attitude" pervades all conscious experience. It is always "directed towards" something such that consciousness is always-already "full" i.e., "about" something. Said in the negative, it is never without an object. This intentional structure (or simply, intentionality) indicates "how" awareness takes aim at its most-meaningful content while excluding equally present perceptions and sensations.

As an example, in conversation, one's intentionality is focused on following, and reflecting upon, what's being said, rather than, say, a bodily discomfort. If one becomes so uncomfortable, this sensation will slowly rise into one's focal point such that it cannot be ignored, placing itself as the front-

and-center issue with regards immanent experience. Finally, even if we know not what consciousness *is*, we do know exactly to what we refer when we mention it and therefore find consciousness at the dichotomy of its "lived familiarity" as "substantial mystery."

To claim that physical things are observer-independent and exist apart from witnessing beings is to commit to **realism**, but it is a self-evident fact that not a single observer has ever known anything "outside" of their consciousness... ever. It is for this very reason that the Indian mystic Sri Aurobindo Ghose remarked that "it is true that there is no such thing as an objective reality independent of consciousness; but at the same time there is a truth in objectivity and it is this, that the reality of things resides in something that is within them and is independent of the interpretation our mind gives to them and of the structures it builds upon its observation. These structures constitute the mind's subjective image or figure of the universe, but the universe and its objects are not a mere image or figure. They are in essence creations of consciousness, but of a consciousness that is one with being, who substance is the substance of Being and whose creations too are of that substance, therefore real. In this view the world cannot be a purely subjective creation of Consciousness; the subjective and objective truth of things are both real, they are two sides of the same Reality."

Although from our vantage point it cannot be *proven*, we are fairly certain that there exists a world external to us, regardless of the fact that modern philosophy — at least since the time of Descartes and later made explicit by Kant and which is now corroborated by recent developments in neuroscience — assures us that we don't know what the world is "in itself." The mind necessarily introduces structures into reality thereby making it conceptually coherent.

Furthermore, we perceive and thus come to know the world as *mediated* by a multiplicity of sense organs, each one of which is fallible and can get it wrong. For instance, we can be misled by a mirage and pursue an ever-fleeting oasis only to succumb to dehydration in the desert. We can fall prey to illusion and "see" something that was never there at all. We are apt to hear the wrong word, or, by including thought as a sense, we can believe in false gods and nevertheless live our lives in light of that erroneous belief. That we can get it wrong means we have a sceptical problem with confirming the veridical determinations that structure the external world. But the positive consequence of that brute fact (i.e., that we can get it wrong) is simultaneously a condition of possibility for our getting it right, which we do, most of the time.

˙ Ghose, A.: *Life Divine.* (1919)

By the Light of Awareness

That our senses mediate or modify the information we receive about our environments compels us to believe that we do not achieve — that our cognitive apparatus does not even *allow* us to achieve — an accurate presentation of the world. Instead, Kant tells us, with space and time serving as intrinsic "forms of intuition," we re-present the external world, in our heads, for ourselves. This forms an ineliminable structure of consciousness. It allows us to see and thereby define "awareness" more vividly, as the **representational structure of intentionality.**[*]

Descartes points out that we know *only* this **intentional structure of representation** — the generative engine that produces internal representation — with *undeniable certainty*. All else is hearsay. If we take this to be true and see that we have only ever known the external world as mediated by our cognitive machinery, filtered, and possibly distorted by it, then we know not what the world is in-itself. What we know of the world, is "really," a representation of it manufactured within our heads. A place where the two meet and are each altered, in some unforeseeable and irrefutable way, by the transaction.

Knowing that we only ever begin our theorizing from within the confines of the mind denotes what the philosopher Markus Gabriel calls "the **modern stance of reflection.**"[†] Understanding it as an insurmountable problem, designates what his contemporary Quentin Meillassoux refers to as our "transparent cage:"[‡] the epistemological straitjacket that locks us into knowing only our representations of the world and not the world itself. The modern stance of reflection is best defined by Franklin Merrell-Wolff's mystical sentence: "I am the pure Light, which by illuminating everything gives to everything existence for me, and except as things exist for me, there is no meaning in predicating existence of them."[§]

To top it off, modern neuroscience backs this up. It considers conscious experience to be synonymous with the brains ability to hallucinate perception and cast its net "outside" of ourselves in the form of a "best guess" hologram. The nervous systems neurons fire sending signals to the brain, which are algorithm processed in parallel and voila! Abracadabra, consciousness of the local environment, perception revealing a sphere of virtual action. All of this is simply to corroborate what Descartes made explicit years ago, mainly, the problem of the veridical determinacy concerning that which is outside of us, the problem of the external world.

---

[*] See the glossary for further notes on this distinction.
[†] Gabriel, M.: *Mythology, Madness, and Laughter.* (2009)
[‡] Meillassoux, Q.: *After Finitude.* A&C Black. (2006)
[§] Merrell-Wolff, F.: *Transformations in Consciousness.* (1995)

The modern stance of reflection signifies the starting point for any honest, metaphysical, or ontological theory as "consciousness" is the only "place" by which things come to be known, and it is within its confines that propositions concerning what lies beyond it can be made. Although we need not adhere to this result, the insight made explicit by the modern stance of reflection also has the unfortunate consequence of making **epistemology** *prima philosophia*. Epistemology concerns itself with the nature of knowledge and is a meta-science as it makes knowledge claims about knowledge. But again, we need to restrict ourselves to it for one aim of this project is to place epistemological objects (i.e., thoughts, beliefs, judgments, etc.) on ontological grounds.

Finally, it is worth quoting Merleau-Ponty at length: "The entire universe of science is constructed upon the lived world, and if we wish to think of science rigorously, to appreciate precisely its sense and its scope, we must first awaken that experience of the world of which science is the second-order expression. Science neither has, nor ever will have the same ontological sense as the perceived world for the simple reason that science is a determination or an explanation of that world."

It is only by the light of awareness that Being becomes known, and even though the modern stance of reflection tells us that mind is all we have immanent access to this does not negate the reality of reality existing apart from and before its advent. Although rational reflection does not give first-order knowledge as does empirical experience, it does grant us second-order knowledge. Knowledge of which we are just as certain. The world, understood naively as matter, exists, and so too does mind. How the world may be structured to arrive at this complimentary situation is the task of part II. But first, let us expand our notion of what is meant by "the world."

 Merleau-Ponty, Maurice.: *Phenomenology of Perception.* (1945)

> "Reality is a dome of many-colored glass, and from our little corner each of us sees a different combination of colors in the kaleidoscope." - Will Durant

CHAPTER II

Spheres of Being | Fields of Sense

External to our consciousness lies *that* which exists apart from us. From the astronomic to the nanoscopic, the sheer number of "natural," physical objects we've come to discover and catalog is immense. We have a word for almost everything, and as our insatiable curiosity for knowledge continues, even the once ineffable slowly comes into view. Add to the store of the already-encountered with the as-yet-unfathomable, and the proliferation of things we might encounter approaches the infinite. Not only this, but the dynamical laws governing our physical universe allow for — in fact, *encourages* — the exponential evolution of its inherent complexity.

A deep link exists between entropy, information, and complexity as it is this efflorescent dynamism that's responsible for the incredibly diverse world of form and Life itself. We stand as living testaments to the fact that somewhere along the universe's arduous evolutionary path, it has led to the genesis of an entity that has access to objects we no longer think of as physical. A "place" where we find consciousness, mind and its properties, underlying matter.

We are unique creatures due to our inimitable form of consciousness. It has allowed us to achieve technological mastery over our local portion of the world. We have arrived here primarily by adhering to a particular view as regards the universe's true nature, that of reductionist **materialism**.

Materialism is the philosophical worldview that sees everything that we are in touch with as a consequence of matter/energy's endless transformation and seeks to *reduce* all things to just that, the interplay of atoms in the void. On the face of it, this view says nothing of — indeed, does not even attempt to explain — how it is that any of its energetic movements, processes, or combinations thereof, *produce* the phenomenon of consciousness: that inalienable aspect of existence that we are never divorced from, and always-already, so intimately *live*. Not only does materialism say nothing of consciousness, but it waves away considerations of most mind-dependant objects like emotions, concepts, and social reality in general.

Spheres of Being / Fields of Sense

Aside from the outer world understood to exist apart from us, as conscious agents we also experience an immeasurable inner life or *that* which originates from within us. By simple introspection, we uncover the reality of an "internal space," a space we find to be as equally unmeasurable as so-called outer space. Not only is our experience of reality both perceptual and conceptual, but part of it, possibly the most meaningful to us wandering and wondering apes, is semantic-emotional. We also "know by heart" and can become swept up in its emotional turmoil. We do not know pride, shame, or love through perception or conception, but through the heart. The "schema of the heart"* denotes the interpersonal, public sphere of human social connections; a domain of being that *generates* our many feelings. And it is in this affective portion of our consciousness that it knows an undeniably real, iron *will*.

    Our emotions can take over our very being, often against our will. In Love, we open ourselves toward the Other, falling into them and their endless depth. In Shame, we see our "self" by the light of an Other (or Others). In Hope, we willfully project desire toward an open and unknown future. And in Despair, we experience the utter futility and emptiness of Hope. Human emotions are indeed powerful, and for most, their existence is without question. "Feelings," even if "emergent," are real, and above and beyond their constituting elements. That is to say, they *do not* reduce to mere chemical reactions or hormones. Acknowledging their reliance on chemical processes for their coming into being, lived-through emotions are as they are in and of and for-themselves. They transcend the chemical soup that facilitates them, becoming a whole, as Aristotle would say, that is *more than* the sum of its parts. Love exists as Love itself; it's no mere "purely abstract concept," nor is it molecular hormones, although those do indeed relate to love's being brought about. To reduce one of the most potent expressions of ourselves to mere concepts or its physical antecedents loses the very "what" of what love, or any emotion for that matter, actually is.

Emotional truths, as well as spiritual truths, cannot be proven by appeal to perception or intellectual reasoning. They can only be known within the relevant domain of experience that bears them out, which again, belongs to the schema of the heart.

We are, through and through, bio-psycho-social beings, and the psyche *in us,* that which is revealed to us throughout and by life experience and simple introspection is as real as it gets. The human psyche is part of and therefore is, reality. Our thoughts are real, and the moments of their transpiring are as real

---

* Steinbock, A.: *Knowing by Heart.* (2021)

as the chemical events that correspond to and underlie them. The human psyche itself has needs and goes through periods of injury and healing. Just as our physical (biological) bodies require certain things for continued health and existence, so too does the inner realm of the mind require nutrition.

Physically, we need food, warmth, water, and sleep to function properly. Bridging the gap into the psyche we also require shelter for it supplies us not only with a place to sleep but a place to *feel* secure. The need to feel secure and be mentally healthy is just that, a psychological *need*. We also need to feel a sense of belonging and that our life has meaning.

Much of what we know as the psyche as it pertains to the interpersonal sphere of people is generated through empathy — a social phenomenon — and by it, we can also suffer moral injury; just ask any decent person coming back from war.

Excuse the following tangent but most of these people woefully regret their actions even though they were coerced or influenced by circumstances beyond their control. High levels of post-traumatic stress disorder among veterans are a sad and terrible fact. In harming others, we harm only ourselves and war never alleviates suffering but only multiplies it. This is precisely why the State and its army's most useful tools for brandishing it are achieved by "manufacturing consent"* and then instilling within its ranks racism toward the Other (whosoever They may be) and a "Band of Brothers" mentality concerning themselves as their own in-group. To get people to kill other people we must first instantiate 'group think' and form our army into a coalition of "brothers." The best way to achieve the necessary level of bonding is through difficult training. When it comes time for war, the State will begin the dehumanization of the Other — we, the Holy (insert group name) must defeat the Unholy (insert group name). One could interchangeably put the terms, state, nation, savages, infidels, imperialists, conquerors, proletariats, bourgeois, capitalists, communists, socialists, liberals, conservatives, democrats, and republicans, into those brackets. Suffice it to say, in all conflicts, the underlying methods and opposing structures are the same.

To return to considerations of mindedness, as previously mentioned, we also carry within us — ushered in by the act of conceiving and accessed by our language-augmented thought processes — the capacity to know many different realms of experience inaccessible to other animals. A sliver of which is a (ratio)nal, abstract-analytic domain where we can follow an algorithmic process to arrive at the truth; where we use mathematical equations to predict the

---

* This is Noam Chomsky and Edward Herman's term for the method and aim of propaganda.

results of physical situations. As is the case with the physical domain, this mathematical sphere of intelligibility, knowable by reason, is structured by facts. All of this is to say that as rational agents we are in touch with a highly precise, abstract but totally real "domain" of existence; a branch of Being populated by numbers and operators, and their endless manipulations and functions.

We refer to this field of timeless, aspatial objects as 'mathematics.' We uncover this "mathematical world" like how an archeologist's dig reveals an already-there prehistorical situation. "The Mandelbrot set," Roger Penrose writes "is not an invention of the human mind: it was a discovery. Like Mount Everest, the Mandelbrot set is just there!"* The mathematical domain is a **"field of sense"**[+] whose elements are timeless, aspatial objects that relate in certain ways. Human reason explores this realm by carefully uncovering stone after stone — by "seeing" more and more mathematical relations as novel objects — thus endlessly revealing the crystalline structure of mathematics as such. Both pure and applied mathematics are equally real, it is just that applied maths overlap in certain ways with the predominant physical field of sense we refer to as "the universe."

Seen in this light, mathematics and propositional logic are but a small part of a larger "conceptual domain" explored wholly by thought itself. The cognitive doors to which begin to creep ajar when an infant begins to acquire language through an empathy-generating process known as 'joint attention.' Through joint attention — that is, the combined focusing on a perceptual object while naming it — infants come to discover the public nature of the social world, and come to see themselves (and others) as autonomous beings, all while revealing their capacity to be individuated and categorized. As such, language plays no small role in the creative acts of intellectual discovery and mathematical understanding.

This "conceptually accessed domain" of information and numbers, rational and mathematical functions and forms, theorems, and logical operations hide "underneath" and mirrors — the perceptually given — material world. It is this loose coupling of the perceived and conceived domains that allows us to manipulate material with greater precision, thereby increasing our technological mastery over it. Consider, it is by understanding, with a precise mathematical description, the electron that has allowed us to create everything subsumed under its name, electronics. This rational-mathematical domain possesses tremendous utility in predicting the future behavior of matter and thus has undeniable truth value. It has a sense of "Being," that is, it *exists* and

---

* Penrose, R.: *The Emperor's New Mind.* (1989)
[+] Gabriel, M.: *Fields of Sense.* Edinburgh University Press. (2015

is undoubtedly *something*. Let us for the time being call this underlying, hidden "layer" of existence the computational, information-theoretic machine-code sliver of reality.

Atop our emotions, our ability to share meaning intersubjectively opens another realm to us. For our capacity to think-in-words and manipulate categorical language allows us to create a vibrant, collectively constructed social reality. Social reality is but another sphere of existence, a branch of Being, that we are in touch with and thereby experience. It is the place where *we make things the case*, a place where we take "somewhat" fictitious ideas — nation-states, institutions, formal rules and regulations, laws, dividends and derivatives, and other financial and social "objects" — and treat them as if they were real. Indeed, they are just as real as numbers, but instead of *uncovering* them, we *construct* them. In forgetting our semantic contribution to these objects — that is, that they *only* attain existential weight only by and through us — they nevertheless seem to have gotten away on us, taking on lives of their own. This is because they are real, even if "made" by us. When seen for what they are, we understand them in the appropriate and relevant context, as real social constructs.

Indeed, social reality is not merely our construction but our production. Its objects have "come out" of us and occupy a place in the "world" (the world as not yet defined *only* in conjunction with the material energetic system known as 'universe') and they therefore also possess "being." That is, social constructions are truly existent *things*. Consider a "larger-order" sociological structure; the headless chimera that is a corporation: while receiving government subsidies, paying little-to-no taxes, and using public infrastructure, a nefarious, multifaceted machine blindly driven by a profit incentive — that is, an insatiable desire to acquire Monopoly money — led by business interests and uncoupled from any moral compass as it is watered-down to its many shareholders, "transnational corporations" are social objects with "rights" and the legal status of "persons."

Individuals, most unknowingly, contribute to these sorts of "larger order" social objects. In late-stage capitalism, the epoch of our age, companies are compelled, by market dynamics alone, to incessantly pursue profit, often while harming the environment and the public in its calamitous metamorphosis. Corporations, under capitalism, are a kind of cancer and as with every type of cancer, it marches itself towards its annihilation, dragging its host (in this case, people, and the earth) into oblivion. And although a CEO may be at the helm of this chimera, if she doesn't pursue its shareholders' interests, she will no longer find herself in that position.

Thus, any responsibility one may wish to place on any individual within a company is inevitably spread among many and the methods by which it operates dispenses entirely with every sort of ethics. A policy structure that

Spheres of Being / Fields of Sense

allows for, nay, encourages endless corruption. Regardless, the point is that social objects of a "larger-order" obtain and that they have their own kind of "force." Just imagine how much oil Shell has moved around the world. These larger-order social "constructs" truly move mountains. Again, not one mind is responsible for the existence of corporations, or even nation-states and governments, but rather, many; these are truly *social* objects.

Although social reality is produced *by us*, such that it takes on the property of "Being," it's not merely imagined by us — a kind of justified true make-believe — but possesses a degree of actuality. To be fair, social reality has a "somewhat fictitious" nature but also has causal power through our belief in its actuality, which in turn influences our behavior thereby making them *real*.

Corporations really are things, as well as all other social objects like laws and regulations. They are maintained through a kind of collective intentionality, the social domain becoming real by rationality's ability to determine it thus and so. However, that social objects exist because we say so, does not also mean that everything humanity produces is noble and good. We play-pretend that money is real and watch as the effort to accrue it leads to a host of negative psychological traits — selfishness and pettiness — in those who horde it.

Under capitalism, the very process that serves to lift a few into unfathomable riches at the same time destroys the capacity of the many to do the same while also leading to the destruction of the very biosphere that sustains it. By our human cognition, we hold these objects to be accurate and real, and in a sense, they are. If you do not pay your taxes, you may wind up in jail, unless of course, you're rich. Thus, despite its reliance on us, the conceptually constructed, semi-fictitious sociological domain is indeed real and part of reality. Arguably, it is the most important aspect of existence concerning human beings. An aspect that cannot truly ground itself, the domain wherein the human person spends most of their time is in this social realm, one whose vaporous and airy abstractions have little weight in "nature."

Of the ways in which we know reality discussed so far — materially, emotionally, rationally, sociologically — each presents themselves to us in different ways, most of which are only available to conscious agents. And to be sure, each mode that reveals novel phenomena is indeed part of reality. We can also see that some of these object-worlds revealed to us are mind-dependant. The material world, as commonly conceived, is mind-independent but is nevertheless "given" to us *sensually* (visually, somatosensorily, etc.), that is, *within* consciousness. All that we grasp, all that we come to know, is "given" to a conscious witness in a self-same way. What I mean to say is that no matter what field of objects a mind can touch and thereby grasp, i.e., the mathematical, physical, emotional, or what have you, it *knows* all of them *in the same way, by bathing it in the light of consciousness itself.* Therefore, as

13

concerns consciousness, we only ever "know" an "appearance" of an object — even if it be "physical" — and never the — unmediated by our cognitive faculties — thing-itself.

Thought and The Infinite

All of this is to say that we live in a complex Reality. Indeed, the world in which we find ourselves embedded has room for conscious awareness and the many objects made available to it, by it, and for it. Given this, a novel conception as to what reality *is* is required. Reality should not be thought of as a "place" or even "world" of (in)determinate totality. Instead, we require an image of existence that honors its nature as Infinite. Thus, as a first attempt to define the indefinable; The Infinite is an *n*-dimensional, crystalline fractal that endlessly proliferates as it slides between its many domains of "substantial" objects.

A while back I mentioned in passing that it is our unique, highly-refined form and faculty of thought — our ability to accurately *conceive* — that allows us to explore and participate in the endlessly efflorescent blossoming that marks the Infinite. And it is this 'image' or picture of Reality that I believe we are embedded. Indeed, Gabriel reminds us that "thought is the medium in which we orient ourselves on our journey through Infinity. It points out the directions in which we should look out for the objects and events that matter to us in pursuing our different goals."

The 'Infinite' is the boundless sea of Being's actuality and possibility and "contains" every one of its indefinitely many domains, or better, fields of sense. Etymologically, the word 'infinite' comes directly from Latin *infinitus* meaning "unbounded, unlimited." Split into its parts, *in* codifies "opposite of" while *finitus* "definite."

Although it exceeds any description, here is a second attempt; the Infinite is a timeless, infinite-dimensional, fractal-like, architectonic edifice that allows for the existence of an unfathomable number of objects, of infinitely differentiated sorts, and each of their own "substance," to exist. An effervescent blossom, the Infinite is a dynamic structure coupling Being as Actuality with Becoming as pure Potential. In its "deepest" layer, it is sutured together by immaterial facts, each of which is a "bit" of information, and acquires its best mathematical "representation" by modeling it upon the infinite-dimensional space (Hilbert space) that underlies quantum theory. A model we shall consider in due time.

Illuminated by our unique form of consciousness as that which allows us to know the Infinite, the master key that unlocks its features is our peculiar *sense of thought.* Thought is a sense in that it is a dimension of experience made available to higher-order forms of consciousness. Like breathing, it goes on with or without our willful involvement. This is why some thoughts happen to us, while others we author.

For instance, anyone who has experienced suicidal ideation has not had those thoughts *on purpose*. Suicidal thoughts are intrusive, rising to the forefront of the mind of the observer often against their will. We no longer refer to them as suicidal thoughts as there exist many types of thoughts. Suicidal ideation is better as ideas are just that, ideas. Contrary to their actuality, they are immanently accepted as 'good' ideas, only to later be reasoned away for what they were, mainly, *terrible ideas.* All of this is to say that ideas are not judgments or beliefs but are their own class or type of thoughts. It bears repeating, thought is like breathing as it is both passive and active. As an organ of sense, it can be driven – willfully piloted – by the organism producing them, or it can let go of them and have them happen to them. Thought as a sense is unlike seeing or hearing as they are strictly passive.

Regardless, the engine of reason – our cognitive machinery and faculty of *thinking* – grants us unrestricted access to the Infinite's many slivered partitions. By engaging our will and thinking volitionally, we bear witness to its finite and infinite portions and through our creative actions, we alter and upset it, producing artefacts. Through thought and will and their necessary coupling to the physically actual – what we think about we bring about.

Only by thinking have we begun to uncover its hidden structures; its logico-mathematical domain being of particular significance as it underlies and overlaps with that physical fragment of it that we call the universe.

Aristotle already points out, the matter belonging to the material world is defined by only three spatial dimensions. That to be a "body," it must be "adorned" with "Longitude, Latitude, and Profundity" (height) and after these three, there is no passing farther, at least, so far as matter is concerned. The physical world of form and function is restricted to this kind of Euclidean space. With regards to this conception of the Infinite, these mark only three of its indefinitely many actual dimensions of being and possibility.

The Infinite is also the zero-point from which consciousness orients itself in its exploration of "worlds," or "spheres of Being" (i.e., the physical, logical, emotional, etc.). The directions, or dimensions, of which are defined, within and of consciousness by a multiplicity of senses, senses that touch, know, and grasp, the various modes of the Infinite. Every isolable object domain, and others we cannot even fathom, forms a field of sense: a contextual frame that allows for its denizen objects to demarcate themselves, exhibit their relations, and be seen to exist. Fields of sense can also be objects in themselves, consider this passage "

Spheres of Being / Fields of Sense

The Infinite is a dome of many-colored glass, an effervescent fractal endlessly proliferating along unfathomable avenues. In ceaseless bloom, it shares a similar structure to the universe's dark energy and its origin in the Big Bang as nowhere and everywhere is its center. All sentient creatures serve as points of access into the Infinite and are also its zero-points of orientation. Without conscious sentience, only the universe would exist. Without thought, we'd have never come to realize that we don't exist in a world but are instead situated in and sailing through the Infinite. Our sense of thought has given us wings, allowing us to take flights of both fancy and facticity, to explore a reality far more complicated than anything we could ever dream up. Indeed, our imagination exists and is but another modality of the Infinite, one more *way* in which it *is*.

Now, if everything that is able to exist, can and does exist, then maybe "existence" isn't the highest genus. That existence need not be understood as the pinnacle of importance.

As regards the ontology of the Infinite, Gabriel's system is *pluralistic* in that it recognizes the reality of a plurality of "Substances." Very crudely, that Mind and Matter exist, are each real, but indissolubly distinct. This plurality allows him to make another astonishing claim, that there is no unified, underlying ground floor out of which all things come. Poetically, he expresses this idea with the statement that "the world does not exist,"[*] and philosophically, with the word-monster "meta-metaphysical nihilism." Basically, he is trying to say that metaphysics, understood as an effort to comprehend *total* reality, has no object. It is to this I object as I believe he is operating with the assumption that mind and matter are indissolubly distinct. An assumption I hope to reveal as an error with a quantum field theoretic understanding of matter and its consequential immaterialization of the Infinite.

The Übermensch has not yet arrived; we are not yet post-human. There exists only a single race, the human race, as all of our differences are ethnic and cultural. We are all equally equipped in the cognitive domain, as we all possess human-grade consciousness. We are capable of language and the possibilities made available by it. Furthermore, as embodied beings we are expressive of an intra-ontology — we express and understand Being from *within* it. Of our somatosensory experience, our body is no mere material object among others. A material object it may be, but known by us intimately from the inside, an inside whose depth is as vast as the infinite beyond it. An inside that has access to indefinitely many other dimensions of Being.

---

[*] Gabriel, M.: *The World does not Exist*. (2013)

Spheres of Being / Fields of Sense

If all of what we can experience, including that which exists beyond our temporal reach and even the parts we construct, form what we call *reality*, it forms a far greater entity than the merely material-energetic system we call the universe. If everything that we are in touch with exists in a certain sense — our thoughts exist in a certain sense, our errors and false beliefs likewise, numbers and logical operations exist in a mathematical sense, geometrically perspectival 'objects' like sunsets and the horizons exist, stars and planets, and Starbucks and Lulu Lemon all exist, each in their own contextual senses, despite their seemingly different origins — then a novel conception of Reality is required: One we shall call the Infinite.

But what of that unifying ground-floor denied by Gabriel? How might we re-establish it?

Crises and the Hard Problem

We can start by asking just what exactly *constitutes* reality. Just what exactly is the fabric of existence? What does the world "come out of?" Apart from consciousness, we know matter and light most vividly. Given this, one might ask... Is the world as we find it merely that, matter and light? If so, what is their true nature, do they share a common ground, and would that ground allow for the eventual emergence of mind? Does this even express the appropriate way of thinking about these things?

Ignoring mind for the moment, we know matter and light rather intimately. They form — at least in some respect — everything we encounter in the external world. As sensed by us, these two disparate entities make up its totality, including our lived bodies. We consider the material world to be real and "objective," that is, we understand that it exists apart from us. But, if physical matter and light make up all that we encounter in the world, where does that leave consciousness and the human psyche's ever-proliferating domains? What of the mind-dependant realm of thought and the interpersonally dependant realm of emotions? Those pillars that serve as the scaffolding of social reality? Just where and how do the mind and its properties fit into the natural order? If mind needs to be accounted for — as it must, given that it's the primary aspect of our existence — is it possible that 'mind' is what is fundamental? Is there mind under matter? And should we look to our best science for an answer?

We have been struggling with this problem — of fitting "mind" into our ordinary conception of nature — for millennia. Yet the problem of consciousness — this "apparently useless flame thrown up by the heat of cerebral commotion"˙ — remains unsolved. The undeniable reality of conscious experience and our incapability to explain how it is generated or *what it even is* forms what philosophers have come to call the "**hard problem of consciousness**." The hard problem asks how is it that inanimate matter organizes itself such that it gives rise to conscious experience? Why does (or should) conscious experience accompany purely natural physical processes? And why are the two so seemingly different in kind?

The philosopher David Chalmers formulated this expression to contrast it with the supposedly "easy" problems of consciousness. These are the mapping of conscious states with particular brain regions and correlating them with cognitive functions and the like. For instance, an 'easy' question would be mapping the prefrontal cortex with our capacity to reason or associate the hippocampus with the felt urgency of fight or flight. These however don't answer, nor even attempt to approach the hard problem. It asks instead: "Why is there conscious experience at all?"

Chalmers put it rather poetically when he remarked: "How does the water of the brain turn into the wine of consciousness?" A poignant fluid metaphor for what is to follow...

Given what we know of the brain as a constellation of electrochemical feedback loops and neural networks processing in parallel, how is it that a cascade of inert synapse firings produces what we experience as 'consciousness'? The directly lived and immediately felt experience of being an embodied observer perceiving a detailed environment is *nothing like* the story we're told from cognitive neuroscience or fundamental physics for that matter.

To refocus the aperture on the question, one might find it easier to frame the hard problem in the most vivid perceptual terms by considering it as a problem of perception; how is it we arrive at a representation or image of the external world? For if we are to believe the standard explanation given by cognitive science, then "where" is this hallucination, this image of the external world? In us or out there? Or is it both?

To recapitulate the bio-logic story of neuroscience; simply, light striking a retina begins a cascade of electro-neural firing that travels down the optic nerve into the visual cortex for processing and presto! the local environment comes into view, re-presented within our mind's eye. One may be inclined to ask, "where" is the image?

In the abysmal blackness that our skulls encapsulate?

˙ Durant, Will.: *The Story of Philosophy.* (1926)

Spheres of Being / Fields of Sense

Are we simply 'projecting' it back out, throwing up a holographic hallucination that we then take to be the case?

Or is the world as we see it both simultaneously inside and outside of ourselves? This not-yet-answerable question is best exemplified by the famous Pixies song *Where is my mind?* To allude to what is to come, it is omnipresent because "consciousness" predates the systems that *access* it. As a property of space, it is everywhere, simultaneously both private *and* public. In a lecture, say, only I am aware of my internal sensations (hunger, headache, or horniness) but we are all aware of the content and evolving meaning of the lecture itself. Every high-school kid has fought off a boner while trying to make sense of the teacher's meaning. That is, there is a "thing," the semantics of that which the professor is referring, and it exists as a graspable referent to all *presently overlapping* minds while at the same time there are things that remain privy only to those who directly experience them. Not every mind need to explicate it, but this does not equate to its not being available to them.

To return to and make an effort toward resolving the hard problem, our inability to find an adequate answer to Chalmers' question stems from two deep philosophical misconceptions. The first is a mistaken conception as to the 'what' of matter and of our seeing it as Alien. Second, despite our immanent intimacy and ability to describe 'what it's like,' we do not know what consciousness *is*. We know exactly what we mean when we speak of 'consciousness,' but the *"what"* of it remains elusive. If we believe this to be the current state of affairs, that the age-old distinction between mind and matter obtains — and of their supposedly irreconcilable and substantial difference — the problem seems to be insoluble. Thankfully, this is not the case.

To entertain a possible answer and unify the two we need to analyze the very blueprint of the cosmos. To grasp the nature of matter, we will consult the physicists and their research into the very fabric of reality — of light and atomic structure — for surely, it's they who have uncovered the most profound insights regarding the problem of material constitution.

To be sure, modern quantum field theory (QFT) places both matter and radiation on equal footing as both are borne of *fields*. The exemplar of which is the magnetic. Take any bar magnet and one can picture a real albeit invisible "field of force" flowing about it in the shape of a torus. An incessant and seemingly seedless torrent of magnetic influence flows out from and returns to the bar. This invisible field and others very much like it form the bedrock of quantum theory and even make up the very solidity of the bar magnet itself.

A caution, despite having uncovered the structure at the heart of atoms, there remains a crisis in physics as physicists cannot agree on the meaning of their most far-reaching theory. The "interpretations" of quantum mechanics proliferate as there in no consensus on what is fundamentally real in the theory. Indeed, the Fock space formalism of the theory makes no commitment to a physical ontology whatsoever, and it could be understood, with equal clarity, as opposing views, that is, as a theory of strange *particles* or field theoretic *waves*. Add to that, the reality and functionality of quantum physics' primary object — the wavefunction — is debated. Is it ontic or epistemic? Is it physically real or a mere mathematical formality, and how does it evolve? Finally, how exactly does a spread-out probability wave of energy-momentum-charge density "collapse" to "become" a particle? Or does it?

No matter which interpretation one finds appealing they will have to accept theoretical predictions and experimental results that testify to phenomena that are far removed from everyday experience. To account for them, the famous "Copenhagen Interpretation" renounces realism and commits itself to postulating only what is mathematically calculable. As such, it is unsatisfactory. Many-Worlds would have one accept infinite worlds with indefinitely many copies of oneself; also unsatisfactory. But Objective Collapse (OC) interpretations have one accept only nonlocal effects, (which are written all over quantum experiments anyways), quantum state, or wavefunction, collapse, (a process implied by Planck's original hypothesis), and some novel dynamics that don't violate special relativity.

Given this, as will be argued in what's to come, an objective-collapse field-theoretic ontology makes the most sense and even allows for a novel model of mind. Against particle theories, the author fails to comprehend how a gravity "particle" could even possibly exist. For gravity is certainly not "made" up of discrete units of spacetime substance. At the same time, he finds no trouble in conceiving of a graviton as the minimal unit of gravitational potential actualized in the form of a quantum, a spherically standing wave, a kind of pulsating well of influence. This is not to say that the graviton is conceptually well-understood, indeed, it is certainly not, however, there are better intuitive pictures as to its nature, and "particle" is not one of them.

Our main contention with OC quantum theories is collapse itself. We know that — prior to measurement — quanta exist as nonlocal, complex waves. Simple interference experiments bare this truth. They also don't "become" particles, those don't exist. Nevertheless, the problem arises when one makes a 'measurement,' when the quantum in question interacts with a detection screen causing the spatially extended wave to collapse into a single location.

This mechanism of collapse is a mysterious process and results in the famous measurement problem with its various resolutions leading to QFT's many interpretations. In fact, one of the two particulate aspects of quanta is that they only ever interact as indivisible units. The second is that they localize to a near-point to do so. This confers upon quanta the notion that they were somehow always-already particles to begin with, but again, this is not the case. They are more simply understood as waves that misbehave.

One might think that 'measurements' have something to do with us as experimentalists and observers, which to be fair, does when we are the ones purposely designing and performing quantum experiments. However, most simply, a measurement is a transfer of energy that limits the state of a quantum, re-localizing or annihilating it altogether and transforming it into another. Measurement is just another name for *interaction*, the events of which are recorded as spacetime points and have happened since the dawn of time. They happen with or without us.

We will have much more to say of QFT's interpretations later, for now, know that they all understand matter to be an emergent phenomenon that arises from a source that is not itself material. A source that also possesses certain bare qualities we would typically ascribe to minds. Also, a quantum mechanical phenomenon, spin, brings to light the exact symmetry of isotropic space and itself possesses a self-referential element. This understanding, derived from special relativity — that space and time have emergent features — coupled to QFT, opens the door to a unique understanding of what consciousness might actually be. Although we will be discussing both quantum physics and consciousness — a taboo combination to begin with — we will not be claiming that quantum physics gives rise to mind or mindedness generally. We must tread lightly here as the combination of these topics is rife with preamble of pseudoscience and the paranormal. Quantum physics — as naively understood in *any* of its interpretations — is about the fluid-like dynamics of fields and the inert interactions of their energies. *There is no 'life' or 'mind' in quantum theory as such.*

However, *modeling* consciousness on the field-view that emerges from a study of QFT will enable us to understand consciousness in far better terms, enabling us to make explicit many of its unspoken characteristics. For instance, the quantum — the fundamental object of QFT — is always-already instantaneously and nonlocally unified across its entire spatial extent. This is also true of conscious agents, as they all embody an instantaneous awareness of their body's spatial extent (proprioception), and one could argue, equally, a spatially extended awareness of their local environment. With respect to *immediately lived experience* — it does not matter how long it takes a signal to traverse the length of our nervous system, or how many light years it took that photon to arrive at the intersectionality of our both arriving-to-meet-there — the felt wholeness of awareness is always-already unified and spatially extended.

"The foot feels the foot when it feels the ground," as the erroneously attributed to the Buddha quote goes. The disparate parts of our body are instantly connected through a nonlocal awareness of said parts. As to the foot and the ground, the two mutually co-imply the existence of the other. Which, to distance this book from the charge of pseudoscience or poor philosophy, I have stated that quantum physics does not *give rise* to minds or mindedness.

However, one must consider that if they desire the unity of reality and realize that both consciousness and quantum physics exist, then the two *must* relate. Although we do not yet know how. This doesn't mean we can bring nonsense into our theory building. Telepathy and telekinesis are easily disputed by simple and honest observation and New Age nonsense needs to be kept at bay, even though, in the later chapters, I will argue, from personal, immanently lived experience, that a rare modality of consciousness exists. A "mode of givenness" called "epiphany" wherein the mind experiences itself as Sublime.

Some crucial insights we shall extract from QFT are that of the immateriality of reality and of how matter, through superposition and entanglement, accumulate to form aggregates of gravitas, standing waves that behave as attractors, that is, transiently stable wells-of-influence that affect their local environments. An atom does this as it curves spacetime to attract others to it, while its orbiting electron has a similar effect on the EM field and causes photons to become attracted or repelled by it. Another preliminary insight is that as quanta superpose and entangle, they become indistinguishable from one another and sum to one. No matter how many quanta one wishes to add together they will always sum to one, that is, form a single unified system. $1 + 1 = 1$. $543 + 789$ when properly entangled $= 1$.

As such, "you" are an individual made up of some 10 to the power of who-knows-what coherently entangled quanta — an endless involution of ever-evolving and interchanging energy — that somehow persists as a unified individual. The dreamer and architect Bucky Fuller used the term "pattern integrity" to refer to and isolate the unit of psychological continuity that persists over time despite having every atom in its body replaced every seven or so years. A "self" that unconsciously dictates its own design. This "pattern integrity" is here to be understood as the wave-functions unity and capacity to remain itself despite however many quanta it involves in its being.

The primary thesis of this work is that consciousness is to be understood as a kind of field intrinsic to the fabric of reality. Like a quantum field, it is an always-already "there" *property* of the timeless, void-like vacuum. This is claimed while also admonishing the fact that consciousness is something above and beyond that description.

We know to consult the physicists for a peek at the architectural design of the cosmos, but who are we to consult about consciousness? Philosophers of course, three of their schools of thought take as their object of study, not Nature, but instead the "sphere of absolute, experiential immanence," i.e., consciousness itself. But, as we shall see, they too are in crisis, as a certain understanding of our relation to that sphere traps us in a "transparent cage" such that we are unable to arrive at any form of realism. In fact, today's most popular branch of philosophy is called "speculative" realism. This is because we arrive at the scene after Immanuel Kant's great contribution to philosophy, a towering figure whom none can ignore, lest they don't mind the charge of being labeled a dogmatic metaphysician. To articulate Kant's insight with regards to our current project, *knowing* as *minded* creatures has it that we are subject to "forms of intuition" that not only structure but possibly distort, all that we are capable of knowing or coming in touch with.

Despite some philosophers believing that consciousness doesn't even exist or that it is at best an illusion, they all (re)cognize that their research begins and ends within the confines of their minds, the only "place" they have ever come to *know* anything. The only "thing" about which they have absolute *certainty*. This being the case, it makes all knowledge claims about the supposedly "real" world suspect. For the time being, we are going to toss those insights aside and approach the world naively, asserting that what we perceive directly is real, despite whatever contribution we as knowers make in bringing to light the objects of our awareness. Consciousness exists, and so does everything it gets us in touch with. Consciousness is real, and so too are the entities it gets us in touch with. That we intersubjectively produce objectivity does not equate to idealism or negate realism. When we get things right, we know reality-as-it-is-in-itself.

Again, what follows is a realistic account of reality, one that includes minds and seeks to make real what we would normally call ideal. We will ruminate about what exists, however, it does apart from and, in some cases, only for us. Even though our shared and personal thoughts about electrons are not themselves electrons, they are indeed *informed ideas*; arrived at only because they truly do speak about real electrons. Even if our concept is only partial, it is nevertheless *determinate enough* to encapsulate its reality. Electrons obtain; our thoughts refer.

That previous, seemingly innocuous sentence hides the deep philosophical problem; correlationism. **Correlationism** is the notion that thought and being are always interrelated such that we never access the world as it is "in-itself" but only ever know a distortion of it engendered by our ineliminable cognitive relation to it. Subject and object constitute one another in an irretractable relation, epistemically locking us into our minds and forming the invisible bars of our "transparent cage." I will endeavor to show that although this is sound philosophical argumentation, by restricting itself to the form of **relative consciousness** that begets its subject/object structure, it misses out on its **transcendental** form and the knowledge provided by it.

Given the state of philosophy today, let me be exceptionally clear about my ontological commitments: the physical world we encounter and the mental world we construct — or encounter, if you prefer, exist and are real — each in their own sense. That our senses can be deceived or that thought can get things wrong and we can subsequently live in error by entertaining as correct false beliefs, does not also mean that we don't most of the time get things right. Again, it is when we get things right that *we know reality as it is in-itself,* as it exists, or would exist, apart from us. We transcend correlationism simply, by observing the obvious fact that something can only hold true when it is true and when thought and being relate in truth, both obtain. That the thought is one thing and its referent another doesn't matter, both are equally existing "things" with regards to total reality.

As for illusions, in the mental realm we produce and encounter, there exist real fictions. "False" images that arise from real thoughts and therefore exist, for real, but *as fictions.* To see my meaning, consider the fictional character Gandalf from the Lord of the Rings. In this literary construction and our collective imagination, he exists. His genesis was as a thought in the mind of Tolkien, a thought that occurred in the Great Chain of Being that is Reality. Gandalf is a wizard, but he cannot take on the form of a human being and cast spells upon us because his existential value is propped up by our imaginations and when confined to this domain, Gandalf can be understood as a real *fiction.*

If we fail to recognize the contexts by which certain objects exist, we will forever remain ontologically confused. Human thought is real; imagination is *real* imagination. That our imagined objects lack substantial form is not a testament to their unreality. Indeed, QFT, when properly understood, paints the whole of Nature — the physical world that begets our preference with dominant reality value — as equally immaterial. Nature and consciousness "are made of such stuff as dreams are made upon." This book seeks to overcome the property dualism of mind and matter by thoroughly analyzing what both "are."

Both the physical and mental — the scientific and manifest images — form the bulwark of the reality that we are in touch with and can thereby *know.* To unify them will require reassessing what we know about each. To put it bluntly, both mind and matter are composed of *fields,* these, in turn, are "carried" by an indescribable yet knowable somewhat. That "knowable somewhat" is, in physics parlance, called the vacuum. Transcendental consciousness knows this for what it is *as* it is. As we'll see, the oceanic ontology of QFT gifts us a powerful model for understanding much of this mystery we live.

There are some things that we make the case but equally as many other things that we don't. Finally, if everything that we are in touch with exists and is real, at least each in their own sense — whether they be real, mental fictions (fantastic thoughts) or real, truth-obtaining thoughts (that apple computers exist) or real, perspectival physical illusions (horizons, gestalts) or real, physical determinants (electrons, photons, fields) — maybe, what we should instead seek is, the Truth. That is, if 'existence' or 'realism' isn't the highest genus, maybe we ought to make an effort to uncover the Holy.

Physicists idolize the "physical" world, that which we find to be actual without us, the ontologically primary realm we call 'Nature.' Philosophers go further by including consciousness in their theory-building and thereby honor the creative, embedded element of human experience. Their object-domain exceeds the bounds of so-restricted 'Nature.' To understand what "reality is" requires a marriage of the two, to take as axiomatic from both disciplines relevant, self-evident truths. For instance, that consciousness exists and cannot simply be explained away. Or that quantum theory must *somehow* — even if we do not yet, at present, know how — relate to it. Only then might we arrive at a coherent conception of the Infinite; of what "this place" actually is.

We have established that anything we encounter or could encounter within the world contains degrees of being. That is, they are all something, and we occupy privileged positions with regard to the Infinite nature of Reality. Despite all of this and curiously enough, it is through our very ignorance of consciousness that we have made our most exceptional technological progress and consequent mastery over the natural world.

With our experimental scientific method and our logico-mathematical, formal-abstract language games, we've generated tremendous knowledge by ignoring the very *thing*-doing-the-knowing. Since the turn of the Enlightenment, scientific orthodoxy has ignored *our access* to reality in favor of upsetting it instead by way of experiments and encoding it into mathematical structure. This undertaking has not only returned deep understanding but has also resulted in a fantastic turn of events, for our efforts to determine what the world is has brought us back full circle. It's as T. S. Elliot mused:

"We shall not cease from exploration
And the end of all our exploring
Will be to arrive where we started
And know the place for the first time."

Elliot's **strange loop** returns us to ourselves, a place where we are beginning to discover mind under matter, at the bottom of the world.

"It is a wrong philosophy of matter which has caused many of the difficulties in the philosophy of mind..." — Bertrand Russell

CHAPTER III

On Matter | Being Known

What is the intrinsic nature of matter? Given that matter is the essential part of everything we consider "natural" and undoubtedly *real*, what is its true nature, and from what is its origin? Furthermore, what is the mechanism responsible for the difference between inert matter and living organisms? That is, what is matter's relation to mind? Is it merely a degree of complexity concerning matter's self-interactions, of its dynamic electro-chemistry and behind-the-scenes algorithmic computability, or is there another "force" animating the living, an *élan vital* if you will? Could that "extra something," that vital impetus, already exist within matter itself, hidden under the inert inertia of things, awaiting its liberation upon achieving the right conditions? Asked differently, does matter harbor within itself the seed of life and consciousness? And if so, what is that seed?

By restricting itself to what it can measure, modern science has generated much positive knowledge and is characterized by a reductionist, materialist metaphysics. Materialism is the substantial metaphysical view that everything that exists or everything that counts as existing - is material by nature, or put differently, reducible to matter/energy and its force-generating interactions. Energy — here used in Heisenberg's sense — is a substance rather than merely the capacity to "do work." Its guiding slogan, that "everything is energy," is oft spoken as a common folk-metaphysic and is rife with mystical connotation. Although they are not, materialism, thanks to Einstein, sees matter and energy as equivalent. As a metaphysic, materialism is our first attempt at a unified image of reality and because of this, it is beautiful, but it is also wrong. It is misguided because of what it leaves out, for where does its instantiation leave Love and Justice? Where does it leave everything humans care about such as social reality? Where does it leave mind-dependant objects?

Corporations, consciousness, and conundrums all exist, but they are certainly *not* energy. They may "come about" or emerge with energy as a precondition and existential necessity, but these immaterial objects are above

---

Russell, Bertrand.: *The Analysis of Mind.* (1921)

and beyond mere energy. Given that *everything human beings are in touch with exists in particular ways*, a better question might be 'what is real?' rather than 'what counts as existing?' But that may be for another time. Regardless, using deductive reasoning as a tool, most believe the problem of consciousness can be reduced to and solved in terms of materialism. However, what is often overlooked is that the very foundation of this metaphysical edifice is missing a pivotal cornerstone as we do not yet possess a proper understanding of *what matter actually is*.

In what follows, I will argue that most of the literature on the topic is littered with a misconception of matter; that many metaphysicians hold a view that assumes an erroneous view of its fundamental constituent! This is because matter's most fundamental property — that of 'being solid,' 'possessing inertia,' or 'having mass' — are not inherent to it but emerge from processes that do not themselves possess anything resembling matter so conceived. Matter/energy is not the most basic substance of reality, but it's quintessential; it's an undeniably necessary and significant piece of the existential puzzle, but it is not primary. Nevertheless, we might learn something by considering how we got to the orthodox view of materialism. In a word, via the explanatory power of physics.

Despite the problem of fitting 'mind' into the natural order, physicists have made the most scientific and technological progress by ignoring it. Instead, they have focused the lens of their understanding not on that which knows the world, but rather, on the material world itself. Their heroes are the early Atomists of Abdera, Leucippus, and Democritus. Their ideas set in motion the materialist reduction and gifted us with a simple (partially correct) way to think about the world.

Their cosmology of "atoms in the void" was arrived at by following a pure intuition that if you cut up the material world into smaller and smaller pieces, you'd end up at some limit — the smallest imaginable unit of substance with an infinite abyss surrounding it. They called this smallest unit an "atom" which comes from the Latin root "atomos," meaning indivisible. Today, we understand this "philosopher's atom" — as smallest unit of substance — as a *quantum*: a unit of vibrating *field* energy. The roots of the word "matter" possess elements meaning mother (mater) and substance (materia).

As it turns out, Democritus' atoms, i.e., particles, do not exist, nor does his abysmal void. As a precursor of what's to come, quantum theory provides a far different view as to the fundamental constituents of reality. Instead, it is not particles or atoms that structure reality but *quantized fields*. Their processes and properties differ significantly from the usual conception of atoms and reveal an opposite image to the Democritean notion of "atoms in the void." Quantum fields paint his "void" as an ethereal ocean of energy and his "atoms" as energy quanta that "get together" and entangle in a novel way, becoming the world of our everyday.

Concerning our constant omission of mind, the world that impresses itself upon our senses is made up of 'solid' objects and is most certainly not conscious, so why consider that-which-is-doing-the-knowing (consciousness) to come to understand what the world is(matter)? Today, we are the inheritors of an entrenched language concerning the philosophy of mind. It's easy to see that science made tremendous progress by analyzing the world with the Atomist's reductive intuition. Galileo, led by this hunch, kickstarted the scientific revolution by attempting to *quantify* phenomena.

Implicit in materialist dogma is the erroneous, albeit intuitive, notion that "tiny bits of hard stuff" i.e., subatomic, dimensionless "point-particles," gather to form matter. Today, the reality of "particles" is debated. As we shall see, one can resolve much of the seemingly paradoxical and counter-intuitive structure of modern quantum physics by abolishing them.[*] The present work presents its foundational edifice without them replacing them with indistinguishable, aggregable *quanta*. Although the Fock space formalism of QFT makes no commitment to a fundamental ontology, one can maintain that its primary constituents are particles, waves in fields, or some combination of both.

So-called "classical" intuition leads to a particulate view where objects have well-defined trajectories and are acted on by forces. This view is wrong. Holding to novel quantum insights leads to a field-theoretic view, and an entirely new conception of reality emerges, one that might be right.[†] The field view also opens the door to taking seriously the physical nature of consciousness. For in it, fundamental energy units remain, but they are characterized as oddly behaving *waves,* elementary units that are most certainly *not* particles.

That the formalism makes no ontological commitment only adds to the confusion as "quantum field theory" and **the standard model of particle physics**" are interchangeable names for the same theory. But as we proceed, we shall develop what I consider to be the only sensical "interpretation" of the theory, that of "objective collapse."

The motivation is to model "consciousness" as both fundamental and physical. By the end of Part I, we will understand what matter is and that quanta (not particles) and their fields are sufficient to form the bedrock of physics. Electrons, photons, quarks, and gluons all exist, but they are not particles; they are waves, and when we understand or accept some of their

---

[*] Hobson, A.: *There are no Particles, there are only Fields.* (2012)
[†] To name a few brilliant books for laymen from a field-view. Hobson, A.: *Tales of the Quantum* - Schmitz, W. Particles, Fields and Forces - Wallace, P. R.: *Paradox Lost: Images of the Quantum* - Brooks, R.: *Fields of Color.*

strange behavior, the theory's paradoxes are lost˙. Let's also keep in mind that consciousness, aside from being immaterial and, therefore, not particulate, is rather, field-like. That is, it's expansive, open-ended, and inviting, with the capacity to entertain discrete elements. But we are getting ahead of ourselves.

Returning to naïve materialism, taking it to its logical conclusion results in the elimination of consciousness altogether, such that some of its adherents believe it to be an illusion. Given that they must account for it in some way, as it is clearly something, what better way than 'explaining it away' as an illusion. But we can refute this non-sensical claim quite simply; by realizing that illusions are things that conscious agents can be aware of, experience, and suffer, but cannot themselves *be*. Our sentient perception and conceptual knowledge of the workings of the world is no mere illusion.

On the face of it, materialism has no place for consciousness, the very entity that uncovers and illuminates the world. The mental gets lost in the reduction or is deemed irrelevant to the reality of the world. Consciousness becomes nothing but a mirage, a vaporous illusion. This ridiculous view is known more precisely as **Eliminative Materialism**. Just as its name suggests, it eliminates consciousness by claiming it to be nothing but an illusion as matter is *thought to be*⁺ the only thing that exists — this, from a camp that still believes matter to be particulate. If one eliminates consciousness, then just where exactly is our knowledge of matter kept? How would "knowing" anything at all exist? Does not the bare fact of knowing anything presuppose a witness's existence, of a sentient consciousness? Any view that explains consciousness away, as irreal or nonexistent, is so obviously misguided that their consideration will not occupy our time.

A materialist metaphysics blinds many to the realities of mind and its forces. In studying the inert inertia of the world, many physicists and scientists conclude that it's lifeless and that the "real" is matter and its dark counterpart, energy and its dark counterpart, and the few fundamental forces. Many see these inanimate processes as the only substantial thing worth investigating as it's the only thing we can accurately model and measure, this "natural" order.

At the same time, somewhat comically, they forget the very entity by which they study the mindless universe, their own minds! A precedent that also bars from their taking seriously the apparent fact that what we know and live as our will is another kind of "force." Will — an affective capacity — in conscious agents allows them to defy gravity, pursue prey, or move a cup across a table,

---

˙ This is an allusion to wonderful book Paradox Lost.
⁺ See what I did there? 😌

all with teleological purpose. Consciousness and its **qualia** are real, fully embedded within reality.

At present, we possess an inadequate understanding of what matter is and the problem of explaining "the what" of mind arises when we attempt to explain it in terms of it. Of course, this leads to an erroneous disassociation of the two on the grounds that they are formed of entirely different types of substance; cognitive and extended.

The philosophical inclusion of consciousness and its functions into our worldview is not anti-science but pro-realist, aiming toward truth. Awareness and its various features and capacities all exist and form an integral part of reality as Infinite. To consciousness, everything exists and is real although not everything is true. From physical atoms to philosophical ones, whether a material object or socially constructed one, everything exists and is real. The "problem" with this view is that it becomes the search for Truth.

Also worthy of consideration when considering consciousness is the oft scorned and ignored experiential domain of mysticism. When properly understood and distanced from New Age nonsense, mysticism forms a fascinating branch of self-discovery and phenomenological analysis as it rigorously analyzes various "modes of givenness" or, more commonly, "altered states" of consciousness. As William James observed: "Our normal waking consciousness, rational consciousness as we call it, is but one special type of consciousness, whilst all about it, parted from it by the filmiest of screens, there lie potential forms of consciousness entirely different."[*]

Phenomenology and psychedelic philosophy analyze these forms, crystalize them, and bring them under the judicative powers of rational discourse. A preliminary insight of which brings to light the fact that the mundane, everyday, default mode of our bearing witness is "relational." Another name would be **relative consciousness**, for this is the standard mode where everything stands in relation to another, where the subject/object structure of total experience is well defined. Interestingly, the word "rational" possesses the root "ratio," which speaks of a quantitative relation between two objects and codifies a measure of their mutual interpenetration and dependence. There is a name too — many names, in fact — for the state wherein subject and object mutually annihilate. Consciousness folds upon itself, revealing itself as its own source. The bare point of awareness merges with and becomes the total contents of consciousness resulting in a state of pure bliss — **Sat Chit Ananda** — the immanent-transcendent state of **Nirvana**,

---

[*] James, William.: *The Varieties of Religious Experience*. (1902)

or **Samadhi**. That might sound woo to the hard-nosed physicist, but to the liberated One who has had the direct experience, there is nothing Greater.}

As will be explained, the reductive, guiding light of materialism has imploded upon itself, leading to its own demise as matter is far from material. In his book and of the title, Nobel laureate Frank Wilczek writes: "A central theme of this book is that the ancient contrast between celestial light and earthy matter has been transcended. In modern physics, there's only one thing, and it's more like the traditional idea of light than the traditional idea of matter. Hence, *The Lightness of Being*."[*]

Wilzcek received a Nobel for his 1973 co-discovery of **asymptotic freedom**, a core concept in **quantum chromodynamics**, the theory that gets to the heart of matter. Later, in the same book, Wilczek also observes: "Matter is not what it appears to be. Its most obvious property — multifariously called resistance to motion, inertia, or mass — can be understood more deeply and in completely different terms. The mass of ordinary matter is the embodied energy of more basic building blocks, themselves lacking mass."

Those 'more basic building blocks' are the quanta of various fields, immaterial energy ripples that metastasize as properties of the vacuum. He explains how in quantum chromodynamics, we get **Mass Without Mass**, (his poetic term) thus illustrating a profound point about the fabric of reality, that at its heart, it's not material. In a word, baseless. Physical, sure, but the age-old adage of materialism that "everything is energy" has resulted in a new ontology; one of quantized fields, properties of space whose processional dynamics are governed by principles laid down in real-time by the algorithmic computation of entangled webs of quantum information.

A new understanding of matter is on the horizon. At the core of which we find our ordinary conception of it turned on its head as we now know that matter arises from primordial processes that are not, nor never were, themselves material. Furthermore, that it is begotten by a novel concept known only to quantum theorists.

You see, a consequence of the symmetry of isotropic space reveals an inevitable process at the very bottom of material things. A dynamic movement gives life through an immaterial substance (energy) and hints toward the very structure of consciousness. It is known as quantum mechanical **spin**: a prespacetime, involutive property of the vacuum that through its becoming does a dichotomic co-emergence of inner and outer — of what it's and what it's not, of what is self and is not-self — arise. Spin also embodies the most elementary form of information as it's a binary phenomenon; it is either [this] or [that], or a superposition of the two and, as such, represents the most basic

---

[*] Wilczek, F.: *The Lightness of Being*. Basic Books. (2009)

qubit. This quantum mechanical phenomenon has elsewhere been referred to as the 'mind-pixel,'* for it's the most basic involution and primordial process that folds back upon itself. In conscious agents that have gained the capacity to refer to themselves — creatures who possess an extra meta-degree of consciousness — this involutive structure is its operant paradigm. What a strange and basic phenomenon to discover at the bottom of the world.

Now, not only is matter not material by nature but intrinsic to it's the very seed of self-reference, an (if not the) quintessential structure of conscious awareness. A strange loop lies underneath the world's material and shows itself when the mind wonders about the nature of mind. We wonder what consciousness is while that very wonder is consciousness itself. Put differently, the question about the very origin of consciousness has consciousness itself as its origin, an involutive self-referential process if there ever was one.

At present, physicalism and scientific reasoning do not explain consciousness and its qualia, nor do they even attempt an offering. Its guiding evolutionary theory — Neo-Darwinism — offers no answer with regards to the driver of evolution. It sees evolution as a blind process, starkly in contrast to the real results of evolution itself, considering whatever it may be, produced sight and sense in the first place. Again, Neo-Darwinism ignores much of the phenomena present to the philosophical phenomenologist. Knowledge exists; deliberate, rational thought, memory, and future prediction and anticipation (protention) exist. Subjectivity exists; self-consciousness and meta-awareness — the awareness of awareness — all exist and are real. Social reality exists; promises, rules and regulations, corporations, and corporatocracies all exist. Emotions of self-givenness exist; pride, shame, hope, desire, and even despair — future-oriented modes of being — exist! Love exists as Love itself.

---

* Hu, Huping., Wu, Maoxin.: *Spin as Primordial Self-Referential Process Driving Quantum Mechanics, Spacetime Dynamics, and Consciousness.* (2003)

> "I found myself above space, time, and causality, and actually sustaining the whole universe by the Light of Consciousness which **I AM**." - Franklin Merrell-Wolff

CHAPTER IV

On Mind | Knowing Being as Thinking Matter

Much has already been said of consciousness, but we may still pose the question: what is the intrinsic nature of mind, and how does it fit into a seemingly mindless universe? Given that it's the only "way" we know of any world at all, of what is its origin? Do complex aggregates of atoms produce consciousness? Are mental phenomena merely an "emergent property" of dynamic chemical interactions? Or could 'mind' be the very thing that solidifies the coherent evolution of matter's complex formation? Could mind (or consciousness more specifically) be something "underneath" the immaterial material of the world, something embedded within matter from the outset that "guides" its formation? Not something just "in us," but something intrinsic to the very nature of reality itself?

Consciousness is the most intimately known facet of our existence while simultaneously being the most inexplicable. At once, we know exactly what we are pointing towards, yet struggle to define and explain it. We do not yet know what it is. As Michel Bitbol has pointed out, "consciousness is not a thing, but it's not nothing either."[*] Despite these curious difficulties, the mental phenomenon of consciousness nevertheless forms the bedrock of how we know a local sliver of the world. The possibility for any knowledge whatsoever presupposes a witness's existence — a bearer of understanding — the nexus that points toward is "the what" that is meant by the term 'consciousness.' And whatever consciousness turns out to be, we know that while consciousness is doing-the-knowing, it also illuminates for itself the very objects it seeks to understand. The immediate primacy of conscious awareness is undeniable, yet because of its inexplicable nature, some consider it an illusion or simply irreal.

As 'consciousness' can be somewhat slippery to define, what is here meant by it is the bare property of *awareness*. We (re)cognize it when we wake from a dreamless sleep and always find ourselves always-already *living it*. That is not

---

[*] Bitbol, M.: *Is Consciousness Primary?* (2008)

to say that we are not conscious while we dream, we certainly are, however, the contents of dreaming consciousness are of a wholly constructed nature. In dreaming, our imaginative minds create a perspectival point of view of the dream world and the entirety of the dream world itself. A veritable phantasmagoria of phenomena, the dreamworld need make little sense to our daily structured lives.

Paralleling the shift from the relative to the transcendental mode of consciousness, we can also attain a kind of meta-consciousness while dreaming, becoming aware that we are dreaming and achieving "lucidity." The result: revelatory knowledge wherein we understand our contribution to the construction of the dream and our awareness of it. Despite the dreamworld's unreality, a feeling of epiphanic bliss accompanies the shift from the "standard," first-order mode of consciousness to the lucid state—a state of pure elation that can persist even after one has awoken. Within the dream, the lucid dreamer becomes an omniscient god unto themselves, and propelled by a feeling of emancipation, many lucid dreamers first attempt to fly, embodying that emotion as physically real despite that it's all taking place within a dream. Sometimes, sadly, the resulting bliss from such a realization can wake the new God from the dream, re-fettering their restraints.

Not only is consciousness simply bare awareness but it can also be defined as the capacity for living beings to possess a degree of knowledge and through that knowing behave intelligently. In this sense, we can ascribe a rudimentary form of awareness to bacteria, plants, and vegetation. A sprouting seedling 'knows' which way up is and will flower when it feels the day begins to shorten. Even more rudimentary, single cells exhibit fight or flight behavior and will b-line across a petri dish directly towards a food source or equivalently evade a toxin. Just because cells, plants, and animals, do not possess the depth of degree of conscious experience that humans do does not mean they are without it. When consciousness is understood as a field, the degree by which it is active relates to the degree by which it is distorted, its amplitude. As we know, human-grade consciousness activates the field most deeply, while a bacterium may only distort it minimally.

Consciousness possesses many invariant structures. That is, it is unified yet formed of various senses, or conceptually separable sensational modalities. Each with a focal point blurred around the edges. We see our environment by situated, perspectival visual perception, and hear a seemingly unbounded auditory expanses. Consciousness is almost always about something, and so has an "aboutness" or intentionality. Pervaded throughout it is a sense of spatial and temporal extension and is accompanied by a volitive will that owes its genesis and felt primacy to a proprioceptive, somatosensory body, the surface of which defining a membrane of metamorphosis.

Time, being its unitary dimension, carries a consistent "aspect of familiarity"˙ such that one finds ease falling into an open future despite the fact anything beyond the protention of time-consciousness is utterly foreign to it. Mentally, we encounter intuitions, ideas, and insights, and cognize rational and irrational, sometimes lingual, sometimes imagistic thoughts. Furthermore, it knows by heart, and we can fall victim to uncontrollable emotions of such gripping depth that they can eclipse all other objects of experience, and moods can set a tone and filter that experience over days, weeks, even months.

Consciousness is known immanently as an experiential Self. A generative structure, the Self is also a kind of "space" wherein something is able to appear and can come to be known. It is the embodied, generative locus of awareness, the always-already, zero-point center from which experience is illuminated and becomes knowable — constructed or otherwise — reality. In seeing this nexus and distilling it into an image that moves through time we arrive at the concept of the historically lived "narrative" self — the memory manufactured ego that we speak of when others ask us "who" we are. Never do we say "witnessing presence itself" even though that's what all of us more immediately and intimately are.

Again, in contradistinction to the memory-made narrative self lives its generative structure, the "experiential" self that generates the pronoun I. Its distillation into an image becoming the self's ego. This ego, this "narrative or autobiographical self," is the story the experiential self tells when asked *who* it is. We never give the honest, self-same answer, that is, that we are each of us the pure light of consciousness itself.

One knows their "experiential self" as the inpouring center of what-it-feels-like to be. From this vantage point, we know an undeniably free will and see its pushing back against the incessant influx of the outer world. The external world ceaselessly streams into us, generating the felt immediacy of being a locus of experience, the very nexus we call "I." There are two ways to understand this eye. For this "I" is indeed "a looking." A looking out, a reaching out beyond itself to meet up at the objects of its perceiving, while simultaneously this reaching out is also a constant consumption, a pouring in that, when restricted to vision, is chromatic light.

This "directedness" of consciousness denotes the subject/object relation and is expressive of how things are "given" to us as observers of reality. There is a directionality to this noesis-noema, subject/object structure of intentionality; a two-way street that moves from within to without and vice versa. We know it's "in to out" direction most intimately as a free will, as the Self we live and know as Home. While the opposite direction, "out to in" is

---

˙ Searle, J, R.: *The Rediscovery of Mind.* MIT Press. (1992)

the weight of the world, the object we know as Alien. Just as we cannot help but hear, or see when our eyes open, we cannot stop the world from imposing its impression upon us. Its weight is best expressed by the poetic statement, "what's bigger than us all is on our hands." To grant this directionality a geometrical representation, we could say that this structure is "horizontal."

In the section on the modern stance of reflection, I referred to consciousness as both the "representational structure of intentionality," and the "intentional structure of representation." Seen in this light, this horizontal 'directionality' of the "representational structure of intentionality" moves from "out-to-in," from Alien to Home, while this self-same avenue heads in the opposite direction, from subject to object, when understood as the "intentional structure of representation."

Meta-curiously, although these forms structure consciousness, it is only by its light that we illuminate them and bring them into focus. In this way, we never "get behind" consciousness to study it as we do other phenomena, for it's always-already the 'that' which is doing the illuminative knowing. If one were to reduce consciousness to its most basic features, they would be just that; illumination and knowing.

One could speak of the unconscious, but that would illuminate little in our discussion, for we are interested in the 'what' of mindedness. As consciousness means awareness or the capacity to become aware of the unlimited domain of objects, whether fictions, illusions, hallucinations, or the real, it *knows* them all equally and in the same way. That is, any and every object that we may come into contact with and thereby grasp, that is, know accurately, is by the light of consciousness itself. Only by entering into awareness are objects — whatsoever their constitution — perceived. So far as consciousness is concerned, the only "thing" it has access to is the *appearance* of an object.

The appearance/reality distinction has it that an appearance, contra reality, is often thought to be a simulacrum of some deeper object. Appearances are only partially determined, while their supposedly real referents are believed to be wholly determined and unknowable. This is not the case, as regards consciousness, *the appearance is the reality* and when an individual or group of people know the truth, they know the objectively real.

ᴑ

Consciousness is irreducible and open-ended — that is, not made up of smaller units — and so in this sense is field-like. It's at once whole, without parts but possesses the capacity to move in the direction of depth. For instance, in relative consciousness — its regular, "mundane" modality — a "self" stands in relation to others; a subject of experience encounters the objects of the world and a sharp division is taken for what it is. But in considering the self and going further into the depths of awareness the self can be lost, becoming annihilated in a transcendental state of union with "the world." In this sense, conscious awareness is capable of moving in the perpendicular direction of depth, by instantiating a "vertical" **mode of givenness**.

The term "mode of givenness" is taken from phenomenological philosophy and denotes the "way" things are "given" to witnessing agents. Physical objects, for instance, are "presented" to us such that we enter into a relationship with them, and we can uncover more about them if we so desire, say, by walking around them or picking them up. Even epistemic concepts as contents of thinking are presented to us in such a fashion. Just as we can toss up a cube to see its previously out-of-view surfaces, so too can we mentally juggle a concept and come to know it better. Presentational objects of this kind, both ontic and epistemic, appear to us as "outside" of ourselves. That is, these objects are "not us," they do not form a fundamental part of our Self, and it is in this relating way that they reveal themselves in a "horizontal" mode of givenness — the two-way street that moves from without to within and vice versa.

Consequently, a "vertical" mode of givenness denotes the movement of how a part of the self relates to the self.

Given that consciousness is indivisible and irreducible — whole from the start — it stands to reason that it is fundamental. There are no "proto" experiences or "monads" that get together to form a larger" consciousness. There is only depth in the degree to how "conscious" something can become. Nevertheless, there are 'units' or 'centers' of conscious experience i.e., living creatures and organisms.

We consider our inner world — the intuited, rational, emotional, mental, and meaningful — to be "subjective," that is, personal and unavailable to others' awareness. By our nature, each of us forms a body, a supposedly closed-off container of conscious awareness, divided and separated from all others. Unique in our isolated loneliness, we bridge the chasm by enunciating words; we impose our breath on the physical world. By carrying meaning through language, we blur the lines between ourselves and come to know a portion of the other.

But what if this view — that we are each isolated in our contained consciousness' — is wrong? What if we are simply wrong about what consciousness is? What if, it's as William James once wrote: "That we with our lives are like islands in the sea, or like trees in the forest. The maple and the pine may whisper to each other with their leaves... But the trees also commingle their roots in the darkness underground, and the islands also hang together through the ocean's bottom. Just so there is a continuum of cosmic consciousness, against which our individuality builds but accidental fences, and into which our several minds plunge as into a mother-sea."˙ Is it possible that what we know as sentience is fundamental from the outset? Something intrinsic to the very fabric of the universe as we find it?

These questions and observations aren't new. Surely, they've been pondered since ancient times. Yet it was only around the time of the Enlightenment that the — somewhat useful but ultimately erroneous — distinction between mind and matter was made explicit. In the early 1600s, Rene Descartes deemed there a dividing partition, thereby leaving us with a language steeped in the dichotomy of *rez extenza* (extended substance, matter) and *rez cogitans* (cognitive substance, mind).

Implicit in Descartes' nomenclature is the fallacious notion that consciousness is without spatial extension as only matter is said to be "extended" and, by proxy, is also without temporal dimensions. This is because he restricts "consciousness" to rational "thought" and sees its faculty as aspatial.

Primitively, one might say that consciousness has four aspects: sensing, feeling, thinking, and doing (known temporally as will). Now, by famously stating "I *think*, therefore I am," Descartes gives the faculty of thought primacy. Because they don't "think" as we do, it was by this misstep that he incorrectly saw animals as unconscious automatons. And although that may be partially true, as most don't employ vivid, language-refined concepts to their thought, this doesn't mean that they don't "think" in ways relevant to their ecological niches. Certainly, the "higher" animals we send to slaughter all show knowledge of their impending doom. This ability or capacity, seen as foresight, is none other than a rudimentary *form* of thought, of seeing what is possible, of what might become the case given what a creature knows of their world.

Regardless, consciousness is wholly *more than* thought. It is the very awareness *underlying it* and also bears witness to the sensations produced by sense-organs as well as the emotions generated by the heart and its interpersonal sphere. The faculty of our thinking is indeed ideal, like numbers, thoughts are without spatial extension. But again, this is not the only faculty of our consciousness, or "dimension of awareness" with which we are in touch. We don't just think with our brains, but feel with our hearts, and perceive with intimate senses. We are also "bodily" beings and so are privy to proprioception — a "feeling" of our body's spatial extension which is itself extended and instantaneously interconnected throughout. For instance, we feel

---

˙ James, William.: *The confidences of a 'psychical researcher'.* (1909)

a kiss on the lips *on the lips,* and should our clumsy dance partner step on our toes, it would be *felt at and by the toes.*

Today, having inherited this language from Descartes, we commonly consider the physical and mental as two insolubly distinct and different kinds of "stuff." Historically, we can see in some of the ancient philosophical systems that primacy is given concerning either matter or mind while often denying the existence of its "opposite." Giving existential weight to only one and not the other leads to polarizing metaphysical views; materialism and **idealism**. We know materialism as the view that only matter, or "the physical," exists, whereas idealism is the view that only the mind, or "the mental" exists. Both ultimately fail.

Idealism has a rich history and in the philosophy of mind is taken as its first steppingstone. After all, it is an *idea*lism. At first glance, most wave idealism away, not realizing it takes flight on the wings of a simple fact: that we have never known anything apart from our own conscious minds. To know anything whatsoever, it must be illuminated by consciousness. Accepting both metaphysical views despite their insoluble distinctness leads to a position called Substance Dualism. A view plagued with explaining how an immaterial nothing moves a material body, an issue that goes by many names like the mind-brain connection, mental causation, philosophical zombies, or ghosts in the machine. A proper understanding of matter resolves this predicament, not to mention how easily we move our bodies about. One could say the answer to the mind-body problem is solved every time we move a finger. But by placing mind and matter on equal footing, as we shall do with a field ontology, the problem resolves itself, for the two are indeed one.

In the *Upanishads*, the immortal texts of ancient India, one will find one of the oldest explications that the two share a common ground. The Upanishads are some of the oldest surviving theo-philosophical documents ever recorded — wherein one will discover a primacy given to consciousness. The nameless authors hold that that which we know as consciousness itself is the fundamental ground of all Being. A compelling assertion to say the least, but one that comes to us on the wings of deep meditative insight. An insight grasped in a non-ordinary and transcendental state (or better, mode) of consciousness. A state modern philosophers have come to call an "epiphanic mode of givenness."[*] An altered state of consciousness wherein the subject-object dichotomy that pervades mundane, relative consciousness is swallowed up by a transcendental dimension of depth. In it, although they commonly come to us from without, it is realized that *the way* in which the phenomena of

---

[*] Steinbock, A.: *Phenomenology & Mysticism.* Indiana University Press. (2007)

our experience are given to us are of our own creation. That we, that our minds, are in some deep sense the whole of reality itself. The result, an epiphany of which is bliss.

The idea that consciousness is the ground of existence is not novel. It should not be confused with the post-modern notion that reality is somehow our construct – a mere human-fabricated abstraction. All while granting that our intersubjective social reality indeed is. Nevertheless, this philosophy was given its most rigorous expression by the somewhat cryptic Hegel and his system of Absolute Idealism. In this system, consciousness as ground is an all-embracing spiritual unity. This author is somewhat sympathetic to Hegel's absolute idealism but is here writing to establish it as a kind of materialistic realism. I do not wish to claim everything is a product or construct of the mind or entertain the panpsychist's view that everything is or possesses a degree of consciousness.

Instead, everything takes part in a universal unfolding that is itself an expression of a deeper underlying *field of consciousness.* We are awash in a sea of consciousness, a universal field that underlies and interpenetrates everything—a field we access and express, each in our own autonomous way, as ourselves. Everything we encounter – whether "material" or immaterial - is an efflorescent expression of this underlying and unified field. A field indicative of the baseless fabric of this vision, one we shall later call the **mindfield**.

In this age of post-truth, alternative facts, and fake news; a return to realism is desperately required. We are caught in the information-overload age of post-post-modernism. The human contribution to what we know about the world has been completely internalized such that we now believe we are all entitled to speak our own truths. A erroneous result that will never get us to the Truth. That the world for me is different from the world for you is trivially true as we are each perspectival individuals and abstract our own "reality-tunnels" peculiar to ourselves and our intellectual and cognitive mastery. Nevertheless, we do so only by sharing a common to all, universal ground floor; a psycho-physical, holographic, quantum field. The human psyche and its constructions are real and form some of its "imaginary" parts. The material-energetic system we call the universe is real and forms the solid, situated, furniture of reality.

It is true that science is a cultural affair, and what we uncover about the nature of the world is formed by our very act of questioning it. But what our probing reveals is abstract facts. Real facts of life that are as they are with or without us. Physics and philosophy are not merely models of how we can understand reality but are ways of understanding reality itself and the various ways in which things actually are. We do not produce facts, as Latour would have it, instead, we produce some while encountering and uncovering others.

Our best theories, despite their incompleteness, do indeed reveal ways in which the world is. The theory is one thing, and that which it accurately overlaps with is another, and the two belong to reality as Infinite.

"An organism has conscious mental states if and only if there is something that it is like to be that organism." - Thomas Nagel

CHAPTER V

What it's Like | A Primer on the Phenomenontology of Consciousness

Consciousness illuminates and knows Being. Externally, as a sliver of physical space seen as an observer's local environment, internally, as an endlessly explorable introspective "space" belonging to the creature accessing and generating it. Moving from the observed to the observer, the immediately lived experience of consciousness is incessantly inviting and always-already open to the world. This openness as modality is one of its many "invariant structures" as it is always and only ever this way. Just as consciousness is always a single, situated, unified impression *extended* in space and time, and *about* something, some of its features never change; these underlying structures are said to be *invariant*.

As to this section's title, *the phenomenontology of consciousness* — phenomenology is both a method and school of thought that examines how the various 'phenomena' of our experience are made available *to and for* experience, of how objects of all sorts (ontic, material, emotional, epistemic, etc.) are 'given' to a cognitive observer. While, as a discipline, ontology considers what is meant by 'existence' itself and thereby examines the very nature of being qua being and wonders what constitutes an "object."

With an ontology of fields of sense, objects form the isolatable elements that belong to a particular field. In somatosensory experience, "hunger" can be considered an "object," one that stands in a relevant relation to say "pain," another "object."

As an attitude and method, phenomenology performs a psychological maneuver called the "epoché." This enables an observer to "suspend judgment" — as well as their "natural attitude" that takes everything it encounters as uninterestingly mundane — to ascertain the invariant structures of consciousness itself. As an example, it is phenomenological analysis that is responsible for isolating the five basic tastes: sweet, salty, sour, bitter, and umami (savory).

It is also responsible for the distinction between **relative** and **transcendental consciousness** as well as **phenomenal** and **access consciousness**.

Roughly speaking, "standard," everyday, mundane, consciousness is filtered by a "subject/object arrangement" such that it can be divided into two halves: what we might call internal and external "space." We know this as *relative* consciousness.

Quite probably due to some evolutionary need, every mind unrelenting accepts the world as it is given, as if it were apart from the creature generating consciousness. This external "half" of awareness is *defined and mostly determined by* that side. It is that portion of what we witness that we could call 'Alien' because it is *not* us. In other words, in a highly determinate way, the environment defines a part of one's total consciousness such that it is always, and only ever, precisely what it is.

As to the other "side," the "internal" aspect of our Knowing Being, we could call it 'Home,' as it is generated in us, by us, as us, and for us. We know and live Home as our "Self," this "side" of consciousness is of such an intimate nature that we identify ourselves with it. At Home, we are creative agents endowed with the energy of a free will and can decide what to think about, what to do, and if we shall succumb to, or fight, innate desires. In this internal sphere, often rising up and against our will, we are subject to visceral emotions and sometimes unwanted feelings and sensations.

Not to wade too deeply into it, but total consciousness is knowing both sides at once while *identifying* oneself with the pole they know as Home. Transcendental consciousness, on the other hand, is characterized by "identifying" with the That which *generates* relative consciousness. In the rarer mode, one does not identify in the same way as begets the relative mode. There is no intentionality aimed toward identity whatsoever, only the blissful and liberating understanding that the One is the All. If it *had* to be phrased in terms of identity, one would say that, in the transcendental state, they know themselves as God.

But what of the "what" of consciousness, just what exactly *is* it? Indeed, as a *thing,* it is almost ineffable as it doesn't seem to be a "thing" at all. Indeed, as Michel Bitbol observes: "Consciousness is not a thing, but it's not nothing either." To further add to its mysterious character, *what it's like* to be conscious shares nothing at all in common with the very objects it illuminates and comes to know. Planets, trees, rocks, numbers, mathematical functions, and logical rules, to name but a few "objects" graspable by human-grade consciousness, are nothing at all like consciousness itself.

Add to that, the furniture of reality and the processes by which it evolves and comes into being is also nothing like "awareness" as such. Even purely mind-dependent objects — like concepts and their relations or purely social objects like rules, regulations, and promises — possess a different "nature" than the very mind that perceives them. Although "the mind" knows both mind-dependent and independent objects *in the same way* the mind itself remains elusive and is never itself known in the same explicit way.

When relegated to relative consciousness, we have here isolated its blind spot, for just as we cannot reflect upon the constituting movement that produces reflection, that is, we cannot think about how thought is generated *while we think*, so too cannot consciousness witness *itself.* Again, *while thinking*, thought cannot turn upon itself, conceptualize its own movement, and witness its own essence. Thought can *retroactively* see that thought cannot see itself in action, but it cannot witness itself in this way *while* it is doing so. In just the same way, *consciousness* cannot objectify itself to become its own object and stand in front of itself. It always-already and no-matter-what hides "underneath" and is yet, at the same time, somehow "above and beyond" everything it knows. It is for these many seemingly paradoxical reasons that consciousness remains the central mystery of our times.

However, *not* restricting consciousness to its relative mode may provide a mechanism that may enable it to overcome its own blind spots. To involute its perceiving eye unto itself requires the initiation of veridical experience and the achievement of transcendental consciousness. For only there will the self become identical to its knowledge and thereby be liberated by it.

In his 1995 paper, *On a confusion about the role of consciousness,* Ned Block made a distinction between phenomenal consciousness and access consciousness. Phenomenal consciousness is vivid perceptual experience, the conscious aspect of what it is like to be in a particular state. While access consciousness refers mainly to thought and cognition as "availability for use in reasoning and rationally guiding speech and action." Reduced to its most basic features, Block's distinction is but another way to differentiate between perception and conception.

Finally, regarding "states" of consciousness, a "mental state is something that it is like to be an organism." This "something that it's like" is always for an observer a plethora of phenomena. Any conscious state is always-already a superposition of awareness in that it is an awareness of a multiplicity of "graspable objects" (somatosensory, cognitive, emotional, social, etc.) "Part" of awareness is devoted to bodily sensation while at the same time another "part" is dedicated to sight while yet another is faithful to sound, listening to the harmony of nature or the hum of human activity. This "multiplicity at once" gains a powerful analogy through an analysis of the quantum's superposition state. They too exist in a state of actualizing *many* of its possible characteristics.

Dimensions of Being | The Self as a Well-of-Knowing & Seamstress of Sense

All any of us have ever known is a world filtered through various sense organs and defined by a unique geo-historical origin. Every human being — every single witnessing presence — from microbe to man, has only ever known their own mind's *representation* of things. Ignorance defines the outskirts of this sphere; it is formed of an unfathomable amount of knowable yet out-of-reach information.

What else can we say of that elusive entity engaged in the very activity of reading this? For one, consciousness is immaterial; it's not something we can bump into or grasp in any way. In this respect, it seems to possess a substantiality that directly opposes the world of matter as commonly encountered and conceived.

We live consciousness as an amalgamation of sensation, a unifying movement that sutures sense. The capacity to 'bear witness' is a mechanism that grasps and sews together various "modes of being in touch with the world." Taken together, these form a single, unified impression or image, always extended in space and flowing through time.

To capture everything that a single human consciousness reveals and contains, in even a single instant, would exceed the capacities of propositional language. A cataloged description would have to include the kaleidoscopic field of vision, the felt being of every inch of their body, every ache, pain, and internal sensation, any overarching or underlying mood, any possible emotional quale, their current desires, and the contents of thought, any scents, sounds, or tastes, and a reference to the very ineffability of consciousness itself.

Let's take this moment to make the case that everything consciousness knows is a sense. In the preceding paragraph, we articulated the many aspects of reality that it is in touch with. Some are made by us, that is, we make *some* things the case, (social institutions, regulations, and laws) while other portions of consciousness are *not* made by us but instead are *presented to* us. This portion is the That which exists apart from us, or that which we don't make the case. Regardless, every "way" in which consciousness can grasp something, it is known to it *as a sense*.

For instance, knowing by way of the heart — to be in touch with various emotions — is to have access to an unbounded sphere of experiencing. A domain made manifest by the public nature of our being, where every emotion is known as a sense. One regarding the way in which something is, in this case, how "the self" *feels*. Knowing with the head, via cognitive rationality, the province of thought is a (possibly higher-order) *sense* made possible by our already complex form of consciousness. Unlike emotions, the contents of thought is not bound to the self but can know of its relations to external reality.

By coherently distilling a phantasmagoric scene and presenting it to a witness, as the witness itself by way of a unity of sense, consciousness may best be understood as a *seamstress of sense* and *well-of-knowing*.

As the seamstress of sense, consciousness is the mechanism that weaves together the raw data received by the various sense organs and looking back on them *knows* them by what they have provided. She is a seamstress not restricted to the five primary senses first identified by Aristotle but incorporates somatosensory embodied being, emotions, thinking, and self-consciousness as further "senses." The braids with which she weaves are the "knowable avenues" that consciousness is capable of grasping. These threads are never empty but are "full" of phenomenal qualities, information, and data.

In line with our grand image of reality as a crystalline fractal of Infinite proportion, we are free to geometrically model these braids as open-ended "dimensions" available to consciousness. Generally speaking, as with the structure of special relativity's unified spacetime, consciousness has three "spatial" dimensions, and one "time" dimension: we think, feel, sense, and do. Its "spatial" dimensions have that aspect because that's how *they are* for us, they are, in a sense, *determinate* Being. Thinking is spatial because it concerns itself with the relations of discrete conceptual objects. Feeling because it concerns itself with the *expansion* and *contraction* of the experiential self as the seat of all-knowing. Sensing because it defines bodily being. Vision, for instance, so far as consciousness is concerned, *is* the environment, its very spatial extent. While its "temporal" dimension of doing is expressive of *indeterminate* Becoming. In conscious experience, will and the arrow of time express one and the same movement. The arrow of which is singularly directed because it signifies the endless registration and computation of mutually interacting and proliferating degrees of freedom.

More than this though, the field of consciousness is not relegated to a four-dimensional arrangement but has available to it *indefinitely many* dimensions in which it may vacillate. Just as the imaginary number line is not relegated to the real, but departs from it, so too does consciousness have access to "dimensions" that lie above and beyond the Euclidean plane of thinking, feeling, and sensing. To see how they are dependent upon but depart from one another consider an "imaginary" number. An imagery number is like a real number — indeed, they are as existentially actual as real numbers are with regards the province of mathematics — but instead has a value in an orthogonal dimension with respect to its real counterpart. The combination of real and imaginary numbers results in complex numbers.

With regards to a physical field theory, when we speak of a field's configuration classically, we assign to every point in space a real number. This number is meant to indicate the value of that field's *determinable property* (say, its mass density or gravitational potential) at that location. Quantum fields use complex numbers to assign values to spacetime points. This is because at any "point," there may be more than one determinable property (or one determinable property that has a value in multiple dimensions) lying — in a similar fashion that separates real and imaginary numbers — in a particular dimension. Furthermore, quanta are characterized by imaginary numbers because they possess both a phase and amplitude, and defining them at a single spacetime point *requires* their use.

In this precise sense, consciousness is a kind of quantum field as it has a "value" in multiple dimensions at once. To see my meaning and further analyze consciousness with the ontology of the Infinite, a "sense" codifies a "direction" or "dimension" that awareness can see, move, and oscillate in. Each of which is unbounded. "Death," Wittgenstein reminds us, "is not an event in life: we do not live to experience death. Our life has no end in the way in which *our visual field has no limits.*"

Sight, along with the other four primary senses, are "directions" available to awareness, each a "dimension" of experience rich in phenomenal qualities (qualia), and degree of depth of complexity. Conscious awareness always has a portion of itself invested in these many variated sense modalities as well as shining some of its light on thought processes, bodily sensations, memory, and the generative apparatus that begets a coherent image of the self itself.

Every sense is without limit in the sense that there is always more to uncover. One might say that the visual field is but a mere portion of the electromagnetic spectrum, or that the human ear only hears between certain frequencies, and you'd be right. But this does not mean that the fields of experiencing them are. There is always more to see beyond the horizon, more kaleidoscopic arrangements of colors in the scope; more complex orchestral symphonies of sound.

As an example of "depth," we can speculate that sight — as a "dimension" belonging to the field of awareness — is qualitatively richer for the mantis shrimp than the human being. We *live* human vision but know by science that it is the combinatory result of three "channels" of color (red, green, and blue). The mantis shrimp, on the other hand, is privy to twelve channels of color and can even detect ultraviolet and polarised light. By this, we believe that its phenomenal consciousness, the world of its vision is profoundly different,

---

* Wittgenstein, L.: *Tractatus Logico-Philosophicus.* (1921)

probably far more vivid — psychedelic even — than that of our own. The visual field of the mantis shrimp, when understood as a field phenomenon, is activated and actualized to a far greater degree than humans are able, to a far greater "depth."

Modeling the mantis shrimp as a quantum system — as we shall later do with all autopoietic, self-constituting organisms — we might say that the amplitude depth of that portion of its wavefunction that deviates and samples the electromagnetic (EM) field and is lived *by it*, as its own visual field of vision, is activated to a greater degree than the human is capable.

All the while, *every creature* endowed with an eye, no matter how complex, samples the self-same *field*. As Schopenhauer rightly observed: "We are only that *one eye* of the world which looks out from all knowing creatures."

In seeing, we all see one and the same "thing," actualized to a greater or lesser degree, for it is the field of consciousness itself. The degree that an organism is capable is dependent only on the complexity of its constituting apparatus. The mantis shrimp's visual system is more complex allows it to unfold and compute more information contained within and regarding it.

Not limited to sight, but relevant to every sense that any and every creature everywhere has ever accomplished and achieved is a unique expression and activation of the great sea of sentience; the timeless, unitary, universal field of consciousness.

The engine of reason — thought — is also a sense, a "direction" for awareness, but is a dimension whose bounds are constantly being surveyed by thought itself. Concepts make the unlike alike and generalize things and thought is conceptually structured. To be expressed, the contents of thought must be distilled into more easily understood referents called words. Stringing them together allows us to articulate a thought. The ability to "see" the syntactical meaning of language is but another sense.

As we shall see in a moment, self-consciousness is also a sense and "direction" made available to complex forms of consciousness, of sense weaving. In keeping with the analogy, the threads with which the seamstress sews are various sense modalities, these are not restricted to Aristotle's primary five. Indeed, by weaving with many threads a more intricate pattern can become manifest. And it is just in this way that the "higher-order" functions that populate human consciousness come about. For the more the seamstress incorporates into her tapestry, the more they solidify and obtain, proliferate and superpose to produce even more "senses," ever more threads. With just three threads — say, sight, sound, and proprioception — she can form a braid; this braid may in turn become a novel sense such as thought or self-consciousness. That is to say, it may require the instantiation and summation of the "right" kinds of sense that will later engender the achievement of higher order forms of awareness.

As a well-of-knowing, consciousness forms a transiently stable attractor that endlessly consumes sense. Like the attractors of chaos theory, its insatiable openness becomes a stable system, allowing it to make sense of incredible amounts of data, of information pooling into it. Consciousness has a kind of gravity as the representational structure of intentionality always remains open to receive more. It inhales the environment and exhales through will.

Now, what of that advanced achievement particular to human awareness — self-consciousness?

It is accomplished when we recognize its time-extended nature while attempting to invert its gaze upon itself. When, by involving memory as a sense, it can present itself to-itself as-itself but *as-past* — it achieves self-consciousness: an awareness of awareness.˙

By capturing this well-of-knowing conceptually and calling it the "self," we retroactively posit its being-there and see it as living-through all forms of experience. By "seeing" the self as the space in which objects may appear, we arrive at self-consciousness and become aware of awareness itself. In a word, we become conscious of consciousness. Said differently, by grasping the zero-point orientation of experience — the self — we unconsciously project and integrate it into the very movement of experience and go on to retain a knowledge of its actuality in all of our doings. In the human being, self-consciousness is a further achievement and novel modality of mind. In perceiving the self, another dimension in the overall field of awareness is opened, a braid that awareness can forever thereafter follow. The self is a substance; it exists.

The achievement of self-consciousness might have never come to pass were we not the language-using, social creatures that we are. But it is not an achievement without negative consequences. Its actualization results in its own disconnection with the world and alienation from nature. This is because embedded within the self is a desire to ground itself in that which it knows most intimately apart from itself, namely nature. And on the face of them, the nature of Nature and consciousness seem inexorably distinct. How could Nature, recognized as inert, lifeless, solid matter — how could atoms in the void — lead to or become conscious? The impossibility of answering that question leads to the alienation of self-consciousness as it seeks to substantialize itself in material reality. But by so doing, it achieves only a

---

˙ For more on this topic, see my article The Eternal Efflorescence of Time at https://sophiasichor.com/time/

disassociation from the very ground it once thought it owed its support and existence.

To reconcile this disparity, we need only dispense with the metaphysic of "atoms in the void" and instead adopt the unspoken metaphysical commitment of modern quantum theory—that of the substantialization of 'nothingness' or empty space. Akin to this, the self witnessed by the involution of consciousness is a Substance, just not one of material/energetic constitution; it is closer to a "property of space" or "characteristic" of the void-vacuum as it has the same substantiality as consciousness itself.

"Where" is my Mind?

Is one's total consciousness a private affair or public, already outside of itself and in and of the world? Put differently, do "minds" extend beyond the boundaries of their generative bodies, or are they confined to within and merely "holographically project" that which is without? The answer is that it is both, the result of *relating* in a way. The phenomenologist Maurice Merleau-Ponty once pointed out that by seeing the eagle and hearing its call, our senses meet up *at the object itself* with an endogenous perspective.

Consciousness is the *simultaneous* bidirectional movement that expresses the dynamic relating of two "objects" — Self and Other. In this way, we can see awareness as taking place *at the locus of relation* and is therefore not merely *in us* or *out there*. True to its quantum origins, awareness is a superposition of being both things simultaneously. So far as we experience the phenomenon in real-time and concerning "the how" of its presentation, or givenness *to us*, and if consciousness is to be taken as total experience such that it includes its own contents, we are always-already very much outside of ourselves.

Again, indicative of the superposable nature of "minds," they may "overlap," that is, the elements that structure one person's experience are there also for others. So far as consciousness is concerned, it *can* be a public affair but need not necessarily be. Suppose multiple minds attend the same event, a concert, in that case, there is a possibility that they will each know *exactly* the same thing, the music, albeit from different perspectives, as each needs to occupy a different location.

But to return to the "where" of consciousness and ascertain its "location," an analogy that honors its luminous quality may be welcoming. It may help us to consider consciousness as the light produced by the flame of a single candle: one that is both the flame itself and the reach of its brightness. The flame as localized body is just as much, if not more than the light it throws off. Light, and consciousness analogically, is substantial in that it exists and *is something* but is nevertheless immaterial. Thus, contrary to Descartes, mental substance is as extended as extended substance.

Now let's consider the standard neuroscientific view as it holds that we capture the being of things on our sense organs (our retinas, eardrums, skin, etc.) and then, via some inexplicable wizardry, re-present it holographically within our brains. Essentially, consciousness is only-ever and always *in us:* within the confines of our craniums rests the totality of the world. A reality we re-present and re-project like a hologram.

Our sense organs receive data from the environment in the form of in-themselves imperceptible impressions. We don't notice photons striking our retinas, soundwaves reverberating upon our eardrums, or molecules settling snuggly into chemoreceptors. We know *only* the conscious experience of their taking place. These interactions are *necessary* for the emergence of consciousness, but it nevertheless remains something "under" *and* "above and beyond" these material relations. Interestingly, every physical sense owes its origin to the oscillatory movements of waves, that is, all environmental stimuli that result in sensory impressions are communicated to an organism via some form of a wave. Sight and sound are apparent, but taste, touch, and smell are less obvious, but rest assured, quantum physics bears the truth of their wavy actuality out as atoms and molecules are themselves geometrically bound standing waves. On this take, consciousness is a frequency-spectrum decoder and analyzer.

What we know as physical sensation is nothing but the interaction of a wave of energy encountering the material boundary layer of our body. At the locus of every sense organ, vibrating waves transfer a particular amount of energy to cause a cascade reaction with a person's nervous system: the causal interaction, its location, and magnitude move on to the brain for processing. Again, the standard neuroscientific view is that the brain *then* processes the signal and hallucinates experience.

On top of this, some claim that we live in the past because it takes time for our nervous systems to define themselves. To take in and make sense of an input signal. But in *conscious experience*, the time it takes for a signal to be processed by the brain doesn't matter. As the fake Buddha quote goes: "The foot feels the foot when it feels the ground."

Indeed, senses can be deceived, such that we can get it wrong. But this does not mean that we do not grasp the objects that provide us with our sense impressions or that we only "get" the result of our brain's causal processing. Our brains are receivers of information that in-themselves share a similar nature to that which they decode.

Regardless, the view is incorrect as it is based on outdated concepts of consciousness and quantum theory. In my view, consciousness is an already-present *field* activated and accessed by organisms. The depth of degree by which it can come online for an entity is relative to the level of its complexity. The way it is activated "in us" is as our "subjectivity," while the way we access it "outside of us" is as objectivity. But this is all taking place upon the self-same field that is consciousness itself, where its contents or "things" take on subjective or objective "coloring" based on *how* they are given. When two people are in the same room, a portion of each's consciousness is directed toward articulating the environment by "activating" the shared field as the "that" which they unreflectively believe to be "external" to themselves but is in actuality "internal" to both. The previous sentence only makes sense if we think of consciousness as an *always-already-there field.*

As an analogy, consider the gravitational effects of two neutron stars. An unavoidable consequence of their being spherical aggregates of energy-momentum is that they both distort the fabric of spacetime such that each is their own generative *cause* of gravity. As with the "horizontal," two-way street that underlies relative consciousness, these neutron stars ceaselessly, and without effort, reach out to pull in, each activating and accessing the self-same metric field that underlies general relativity. They do not each produce their *own* gravitational fields but instead, each influences the very same, and only, metric field. Should these two stars come close enough together, their effect on the metric field would — through their superposable nature as waves that distort their host medium — sum. That is, at the places where they overlap, the gravitational potential would increase as a greater depth of deviation concerning the field's "lowest energy" value.

Regardless, the neuroscientific approach to consciousness is lacking because it writes it off as a mere hallucination. Understanding neuroscience as embedded within a false paradigm of materialism tells us why. For the most part, the neuroscientist has been prioritizing a false characterization — classically conceived — matter as primal being resulting in a disingenuous understanding of *both* matter and mind. Taken neuroscientifically, (or with the doctrine of materialism as its guiding light) there exists an insurmountable divide cleaving the ghost from its machine, the immaterial consciousness from its supposedly "material" substrate.

Luckily for us, quantum theory unveils the true nature of matter as lying in stark contrast, in fact, in direct opposition to, its commonly conceived understanding. In a word, matter is not solid, not even material, but instead arises from the movements of field-facilitated, pure energy. As understood by our most accurate and deep science, the ground of reality is made of self-similar *stuff,* the same that precipitates our dreams. What we know intimately as our personal consciousness is but a mere sliver cut from an immense sea of sentience. The cloth of consciousness is an ocean of awareness that every self-contained, autopoietic system participates in, actualizes, and activates by simple virtue of their being-wherever-it-is-that-they-are. Living organisms, even those as simple as prokaryotic cells, are rooted within this Timeless, indiscriminate Somewhat.

Now to speculate. What if, more than a field, consciousness forms the very ground of all being? Not in the sense that human consciousness constructs or somehow creates and holds up the Infinite, or in the panpsychic sense that every actual "thing" is in some sense conscious, but rather that consciousness is, and always has been, a timelessly existent unified substrate, one of which we, and all living creatures' *sample*. Later, an insight we will borrow from QFT is that of the relationship between a field and its quantum – the medium and the kinds of waves it accommodates – as they are expressions of one and the same *thing*. Likewise, what if living beings are – each in their own way, due to their spatio-temporal location and historical genesis – precisely the part of the field they occupy?

Also, what if this field of awareness possesses indefinite dimensions, some of which we know as particular senses? That its differentiable dimensions *really are* coordinating directions that help to define and anchor individuated minds as pockets of sentience via their senses. Might there simultaneously co-exist, in the very same "space," deeper possibilities for the expansion of consciousness? Potentialities that are there, just like the Mandelbrot set, but remain undiscovered because we, or other minded beings, have not yet evolved the relevant sense organs, or combinations thereof, to unfold and extricate them; because the seamstress has not yet woven the relevant threads into a novel, intricate braid.

The form and structure of consciousness is a well-of-knowing whose gravity is acceptance; it is endlessly inviting, an insatiable consumer, and a unifying weaver of sense. Its illumination does not discriminate and transmutes all data into witnessed Being. It is always-already open to receive, and as predominately understood – most vividly through sight and sound – its extension bleeds into a dynamic and living environment.

In the human being, the evolution of consciousness has reached its as-of-yet most refined form. I say this not to diminish the mental states of other life forms but to accentuate the profound difference that humans are. Our sense of thought – an exquisite faculty of rationality capable of reaching across eons of space and time such that it terminates its sense-making by witnessing their incomprehensible origin – has gifted us language and resulting social reality. Not only this but thought, as an organ of sense, allows us to navigate the timeless waters of the Infinite. Finally, its coupling to self-consciousness may form the braid of senses that grants us our unique ability to treat reality as if it were made of clay.

As children, teachers and parents often tell us that we can do anything that we put our minds to. But they almost always forget the caveat: that it must remain *within reason.* That is to say, not only are there physical restrictions but thought itself recognizes them as its own actualizable boundaries.

The evolution of thought and consciousness is not yet complete, as we have not yet discovered (or produced) the keys that unlock its many other, possibly inconceivable, dimensions as doors. Novel sense organs, or novel combinations thereof, are needed. Despite not currently being equipped with them, this does not limit us from knowing about them. Even if we may not ever know what some senses are like, we observe other creatures' sensing differently and thereby occupying an umwelt different from our own. For instance, the echolocation of dolphins and bats, the colorful world inhabited by the mantis shrimp, and the magneto-sense possessed by sharks mark just a few further "dimensions" that consciousness can pursue.

Unbeknownst but sadly for them, possibly most importantly, non-ordinary states of consciousness exist. Although rare and directly experienced by only a few, they provide an unshakable form of incommunicable knowledge. This situation, of not being able to place into words the ineffable nature of what one has come to know was put best by Franklin Merrell-Wolff: "The psychological, epistemological, and logical hackers may tear to pieces the formal garments of systems like those of Spinoza and Hegel all they please and still never reach at any point the inner authority upon which those systems rest. This is so because philosophers like Spinoza and Hegel know what they know despite the defects of their formulations and all the attacks of lesser thinkers. One who has been There is not moved by a mountain of denials from those who have not been There."

So, just "What" is Consciousness?

That 'illuminative knowing,' that we live as consciousness itself, is a unified *field*: an immaterial, mental medium of ubiquitous sentience. It is a field not unlike those discovered by Faraday. The electric and magnetic fields that permeate extension as properties of "mere" space.

With this preliminary exegesis on consciousness coming to a close, we are in a position to make a few more claims regarding its nature — even if some may seem unwarranted, speculative, or even downright crazy.

One, consciousness is immaterial and is of such a primary nature that nothing can get "behind" it or can come before it. As a "thing," it is timeless and therefore possesses no origin; the moment of its genesis does not exist; it never needed to occur. Conceding that, whatever "it" turns out to be, "it" had to have "been there" at the beginning of time. In this vein, it is best to think of "consciousness" as an aspect or quality of timeless nothingness, a property or characteristic of an eternally empty Space. Understanding it in this way, we can speculate that it did not come into being with the advent of life and is, therefore, no cosmic latecomer. "Life" represents its acquisition of form and is expressive of its activation. Now embodied, consciousness can become a subtle gravitas that guides evolution much like the philosopher of time Henri Bergson had in mind with his Masterpiece *Creative Evolution.*

Two, despite the diversity and everchanging nature of its contents, consciousness is a unified "object" smeared over a small window of time. It is always known as a single unified impression whose movement possesses a degree of temporal width. That is, it retains in the present that which has just-past while projecting itself into about-to-happen.˙

Three, the various senses available to it are the dimensions in which it oscillates or moves and as such has a degree of existential purchase within them. An amalgamation of sensation, if consciousness were an empty void, its "force" would be a kind of gravity that solely consumes and unites sense. But just as gravity not only pulls but reaches out in the opposite direction — an incessant consumption which is at the same time a grasping for purchase — so too does consciousness consume and exude. A well of want, what at first seems to possess only a singular direction is, in actuality, a two-way street. To activate this opposing "force" in consciousness and know it as free will requires only the acquisition of form. An insatiable pool of perception, the primary senses supplying consciousness come from without, from what is Alien, but equal and opposite to that flow is its fountain; generated from what is Home, the endlessly blossoming flower we know and live as will.

Four, as an extended and immaterial property of space, consciousness is as much "in us" as it is "out there." This is why when friends share an experience, they know *exactly* the same thing. When a group as a unit feels the mood of a room, it is due to the public nature of consciousness and of how the individual minds comprising the event overlap in the field of consciousness, of how they *superpose.* The perceptual portion of individual minds is perspectival, but what it is able to be known by them is not. Public matters are publicly knowable. Furthermore, if two or more minds grasp the Truth, such as $2 + 2 = 4$, they know one and the same thing. Not each knows their own version of the equation, but each knows one and the same equation. It is like our neutron stars that each affect one and the same spacetime.

We know one another far more intimately than we admit. This is because the field of knowing is not of our own creation but is instead the most common dimension that we all inhabit, the sea in which we are all swimming.

˙ Again, see my article The Eternal Efflorescence of Time at https://sophiasichor.com/time/

Just think of what "sphere of being" is most cared about by humans. Of course, it is the one that it has the most influence in creating; it is the socio-emotional sphere. Made what it is by the interpersonal establishment and public nature of our being. We care far more about human established drama: on a sociological scale things like economic depressions, falling stock markets, and geo-political turmoil, steal our attention while on a more interpersonal scale things like broken promises, failed relationships, and in-group out-group dynamics all come before questions of our ultimate place in Nature.

Finally, we have said much of consciousness, trying to see its light. Now, given all of it, what might a perusal of QFT reveal with respect to it? Well, nothing really. However, although the blueprint of the cosmos as elucidated by QFT says nothing whatsoever about consciousness, an examination of its ontological commitments may still unveil insights with regards to the very mind that contemplates it.

Consciousness is a massless, multidimensional field. We perceive and know *some* of its dimensions as senses. Its force is a kind of gravity that sutures sense.

Consciousness is a property of space (being aware of) and has a value of degree spread over the objects in the total environment. The coffee cup of which you and I are aware is one and the same cup and consciousness. Awareness superposes over the object, "doubling" its existential weight by our shared opening towards it. This doesn't make the coffee cup real or ideal, simply and truly *seen*, witnessed by the light that is consciousness itself.

When we are together and both perceive the same object (whether ontic or epistemic, whether perceptual or conceptual) we know one and the same *actual* thing.

Part II

The Baseless Fabric of this Vision

"What is that which eternally is, which has no origin? And what is that which arises and passes away, but in truth never is?" - Plato, *Timaeus*

CHAPTER VI

Immateriality at the Core of Quantum Theory

What is the true nature of the world as it exists apart from us? Beyond our language games, conceptual schemes, and social constructions, what is the world as it is in-itself? How does it operate, and what are the mechanisms by which it came into being? Is 'everything' simply a material-energetic system evolving based on a few simple physical laws? A ray or eigenvector in Hilbert space? (For simplicity's sake when you read "eigen," simply think "a range of.")

Although highly refined, these kinds of questions have guided humanity's curiosity for millennia, and the physical sciences have long sought to answer them. More accurately, the discourse of physics has long struggled to grasp the inner workings of matter, the world's (ultimately inanimate) substrate, and understand its static and dynamic processes.

Our most profound, well-tested theories form our Core Theory*, an immense body of experimentally verified knowledge. Suppose the basic operating principles of the universe are timeless. If that is the case, we can rest assured our Core Theory will remain at least somewhat correct until even time itself no longer possesses any sense. It will remain true even in light of new evidence and information. For instance, even if it turns out that gravity is quantumly entangled qubits of information, general relativity will forever remain correct as a limiting case and relegated to its sphere of intelligibility, just as Newtonian gravity remains a limiting case within general relativity. Newtonian gravity was not the correct theory of gravity, but it's not entirely incorrect either.

Our Core Theory embodies our most profound understanding of the world. From the chaotic science of complexity to the self-organizing systems of biology to thermodynamics. A universe we have woken up to find ourselves thrown into. A time and place no one asked for, yet here we are. Different

---

* Frank Wilczek unites humanity's most significant physical theories and calls it "Core Theory," as neither QFT or the "Standard Model" includes General Relativity, Chaos theory, thermodynamics, etc...

branches of the Core Theory seek to explain some aspect of the world as it's given to us, as it were, without us. A difficult task, given our embedded nature and constant causal contact with it—our mere agency altering and upsetting it with our every move. A fact we will disregard for the time being.

Humanity's Core Theory brings into the light the most buried truths we have collectively uncovered concerning the fabric of this vision, its properties, operations, and functions. At its core, it explains everything we naively understand to be real — the very nature of light and matter and four fundamental "forces" — in terms of warps and waves of information-carrying energy in fields.

The universe as a material energetic system can be viewed through two lenses of our augmented eyes revealing two axiomatic modes of our inquisition. We stand as giants and manipulate light to bring all matter and form into focus such that it all that looks tiny to us. In actuality, we know planets, stars, and galaxies are far larger than us, but it's rather curious that due to our visual perspective, even the largest objects in the universe appear to us as no larger than a dust-mite. Through visual metaphor alone, we reveal to ourselves how we try and grasp the nature of reality.

The understanding brought about by a microscope reveals a Democritean desire to reduce the world to its constituent bits. The understanding brought about by the telescope reveals larger-order wholes that somehow must emerge from the simple laws that govern mereological sums. A particle physicist is a reductionist while a condensed matter physicist is an emergentist. Naïvely, we believe that what these two groups study captures everything that exists, but the moment we turn to philosophy our world swells. For not only does this seemingly boundless material-energetic system exist, but so too does truth and facts, lies and deceit, feelings and emotions, ethics, and morals. So, "the world" when pondered by a philosopher is far larger than that of those who already reduce their "world" to the universe. In fact, in philosophy 'the world as such' is already a problematic concept.

Reductionism and emergentism broadly capture the how we analyze our Core Theory, each of which guides our search for insight. Behind every piece of knowledge it contains, lies two undeniable and fundamental assumptions that the world — that great "unknown something" — must possess as a bare minimum, a 'being' and a 'doing.' Put differently, a 'whatness' or ontology, and a relational-causal structure. We analyze these threads as operant principles of the world, two "parts" that are each toned by intuition in the understanding.

Firstly, toned by spatial intuition, we know the world in a fundamental-analytic manner. In this regard, we are concerned with the "what" of the world, its fundamental ontology, and the theories that get to the bottom of what reality is in-itself, the supposedly "material," physical universe that exists "out there."

Immateriality at the Core of Quantum Theory

In a word, via reductionism, we find what is *fundamental* about 'being' through *analytical* equations.

Intellectual cognition operates rationally and owes its development to — and is therefore toned by — our intimacy with space. It is by the extensity of space coupled with bearing witness to causal forces between distinct solid elements that endows human agents with the notion of divisibility.˙ Thus liberated, the intellect uses the syntactical structure of language to categorize and divvy up the world into pieces, abstract symbols, math functions, and logical units and their operations; concepts basically. The entire domain of the discrete reveals itself to conscious agents analytically, literally (ratio)nally! It should be no wonder that the ultimate computation machine — the binary information processor that is the human brain — operates, both physically and psychologically, digitally.

Secondly, and toned by a temporal intuition, we know the world as a kind of evolving dynamism; this thread is complex-algorithmic.⁺ The complex-algorithmic concerns itself with the chaotic sciences of becoming, the proliferation of information processing, quantum computation, and the evolution and autopoietic formation of DNA-encrypted entropy harnessing, living organisms. Basically, complex systems. The complex-algorithmic order reveals a type of proliferating computation underlying the temporal mode of causality. A process that leads to the higher orders of complex objects, mereological holons, and relational processes. Life itself is the quintessential example and pinnacle of said dynamism. In a word, via emergence, we capture what is *complex* about 'doing' through *algorithmic* equations.

Motivated by the insatiable desire to understand the intrinsic nature of Nature — of matter, energy, light, and the physical "forces" that keep everything glued together — we will begin our inquiry by examining the fundamental-analytic "whatness" of the fabric of reality. Keeping in mind that "underneath" this domain of the visible lies the invisible physical laws and rules for its coherent becoming; the info-immaterial, complex-algorithmic computational "layer" of reality. This information-as-substance portion of reality contains a discretizing function and serves as its 'machine-code' bookkeeping layer. It is designed (or destined) to keep the world coherent. All of this is to say, these two "modes" of universal Being — the fundamental-analytic and complex-algorithmic — are different aspects of one and the same fabric.

˙ See the works of Henri Bergson for more.
⁺ This insightful division is owed to Glattfelder, James, B.: *Information–Consciousness–Reality*. Springer. (2019)

Immateriality at the Core of Quantum Theory

⛢

Of the fundamental-analytic, the two equation-based experimentally verified theories that get to the bottom of the "what" of the world are quantum field theory (QFT) and **General Relativity** (GR). QFT is often thought of as the theory of the subatomic microcosm, while GR is a theory of gravitation confined to the astronomical scale as it lacks the precision to model intergalactic dark matter. And while this is partially true, they are more deeply theories of the very tapestry of the universe, of its very fabric. Despite their incompatibilities implying incompleteness, when woven together, each form an eternal golden braid that weaves together the universe's very fabric.

As to the unity of reality, the philosopher-physicist David Bohm has so succinctly pointed out that both quantum and relativity theory — each in their own way — imply the unity of the whole of the cosmos.[*] That is, they are each a testament to the view that the universe is an incessant flux, an efflorescent globally transforming, undivided and unbroken *whole*. To bring this insight into greater relief, we shall have to visit the nature of holography and pull clues from its study to form a decent "picture" as to the nature of reality and its various multifaceted temporal movements.

QFT and GR are both *field* theories, and they encompass our deepest knowledge about the world as it exists apart from us. Given that I'm writing from the insight that consciousness is fundamental, we are granted a way to envision what consciousness might be: a universal field with its own degrees of freedom and depth of deviation.

In physics, a **field** is described as a mathematical function over space and time and in just this way is often considered irreal or at best a mathematical fiction. However, fields are real, observer-independent actualities and as such are best intuited as *properties of* space whose organization changes over time. That is, fields codify a capacity for space to do or exhibit a certain behavior. Fields are essentially bare qualities of the quantum foam; the non-void void-vacuum.

There are two ways in which we may refer to a field: as its *property* or function or in terms of its *configuration*. The first codify what a field is (its primal property or determinable aspect) while the second assay its current global status.

In what follows, I will mostly be using the term 'field' to refer to its functional definition as a property of space. In this sense, it codifies the fact

---

[*] Bohm, David.: *Wholeness and the Implicate Order*. Routledge. (1980)

that there exist determinable aspects of the environment, that is, the 'what' about a specific field (i.e., the electric or magnetic vector, gravitational potential, mass-density term, or what have you). Secondly, a field may also be defined by its configuration in actual or probabilistic expectation values.

Laid out classically, a field *configuration* assigns a mathematical object — a real number to indicate an actual value — to every point in spacetime. Taken together, the field's configuration is the sum-total of these values. In this sense, one can model an area's temperature as a classical scalar field. At every point in spacetime, a single number indicates the value of the temperature. Or similarly, consider your present environment as a mass-density field where every point indicates just that. For instance, with regards to mass-density in particular, that different physical objects are made of different atoms, each with their respective "weights," the field configuration would hold differing values for differing types of substance such as metal or plastic.

The quantum analog does away with exact values by replacing the representing mathematical objects, real numbers, with an equally mathematical object and corresponding physical referent, an ***operator***. This maneuver — of replacing actual values with an "eigenvalue spectrum" of values, that is, an operator — is called **first quantization** and by it we see that quantum fields are "operator-valued." Correspondingly, there exists a quantum-mechanical operator for every classically discoverable property a system may have such as an energy, momentum, and position operator.

Operators serve the function of determining a classical property's "eigenvalue spectrum." As a working definition, an operator possesses a "noncommutative structure" and is itself a function over space that maps onto other spaces, a linear map of vectors to vectors in a **Hilbert space**.

Hilbert space is a higher dimensional framework that allows for greater precision with which to represent quantum systems. It is a **vector space** with surplus structure that serves the purpose of defining a metric or length. A vector space is just that, a space whose points are multivalued such that they have both direction and magnitude. Every dimension in a Hilbert space corresponds to a determinate state of some quantum system's potential observables. That the values of these observables, such as position or momentum, are infinite, Hilbert space may, but need not necessarily be, infinite-dimensional.

Most physicists see Hilbert space as irreal, as only a mathematical construct, but I do not agree. For me, I understand Hilbert space to represent the actual ground of the universe for it is infinite in infinitely many ways. Every quantum system is an infinity unto itself nested in a still larger Infinite. Furthermore, many of its possible dimensions need not only be "physical" but "imaginary," marking the avenues upon which conscious awareness oscillates, but that is for much later.

Immateriality at the Core of Quantum Theory

Armed with the preceding discussion, we are now able to technically envision a physically actual quantum field. To approximate an operator at every mathematically describable point in space, one must visualize an array of tiny quantum-mechanical harmonic oscillators capable of being in and expressing a spectrum of relevant values. A low hum, incessantly buzzing, each point in space can be thought of as "realizing" infinitely many omnidirectional springs that are each connected to every one of their neighbors. Each "spring" will have a relevant malleability and when one is jostled it will affect the points beside it. When a wave moves through the field, it's the cohesion of these "springs" that form the field's elasticity, which in turn dictates how easily a wave will flow through it.

As the wave moves through the field, its wavefront lifts a spring, which lifts the next, which lifts the next, and so on. The full height a spring stretches is the amplitude of the wave. We should be careful of this picture; we should also *not* picture a quantum field as a kind of coil mattress. Every mathematically defined point is only a kind of nothingness or bare potentiality. Further to this point, one might argue that the mathematically abstract pursuit of dimensionless points in space is an endless endeavor that can only culminate in an infinite regress... but I digress.

As for the "operator-valued" quantum field, that they possess an eigenvalue spectrum of values — rather than the definite values that characterize classical fields — calls us to envision the field itself as a kind of indeterminate "foam." Acknowledging this inherent indeterminacy is at the heart of quantum theory's many "interpretations." Although the formalism leaves open any commitment to a fundamental ontology, as we know not of what this foam is made, i.e., particle or fields, I will be making assertions that we do know to what ontology it points; one of waves in fields, for it is only when one adds the appropriate quantum of energy to the field do the operators take on "actual" values. To be fair, that may not be entirely correct but what is true is that both quanta and fields are necessary to describe the phenomena.

Furthermore, only a fundamental field view has the explanatory power to model nature and the very mind that contemplates it.

Fields are not things, yet they are not "nothings" either. Later, in thinking of or *modeling* consciousness as a field, we will gain significant insight into the nature of Nature.

Despite their "airy" and ethereal character, fields are not simply properties of space, but are real physical "objects," — that is, they are "things" in and of themselves — and are present in the vacuum even if there is no energy present to activate them. In fact, there is no such thing as a "no energy" state when investigating the vacuum in QFT, only a "lowest energy" state. As such, empty space cannot and does not exist, even if the something that is there — quantized fields — is eerily close to being a nothing.

As far as testable theories go, QFT and GR form the jewels that adorn humanity's crown of — experimentally verifiable — reason. As such, they are the harbingers of a world-shifting view. Indeed, QFT "paints a picture in which solid matter dissolves away, to be replaced by weird excitations and vibrations of invisible field energy. In this theory, little distinction remains between material substance and apparent empty space, which itself seethes with ephemeral quantum activity."[*] While GR transforms spacetime into a gravity-generating, fluidlike, warpable, and malleable manifold.

QFT is the child formed from marrying **Quantum Mechanics** (QM) and **Special Relativity** (SR). A combination that spells doom for a particle ontology as particles cannot satisfy the constraints placed on them by both theories.

The essential nature of QFT's formal structure is best captured through linear algebra and Dirac's *state vector* treatment. However, for our intents and purposes, we are going to stick with the more easily intuited *wavefunction* picture of Schrödinger et De Broglie. State vectors are something with which most people have no familiarity, waves, on the other hand, are quite familiar to everyone and the cosmo-oceanic picture it paints is an intuitive one.

At the same time, GR is a stand-alone, evolved behemoth of SR, differential geometry, and Newtonian gravity. GR explains gravity as energy-momentum's capacity to geometrodynamically affect the malleability of spacetime, the resulting curvature of which we know as the gravitational force. For completeness' sake, GR can be modeled with equal accuracy and explanatory power as either "spacetime curvature" or as a classic "metric field."

Both are continuous field theories, yet QFT is "quantized," possessing discrete elements called quanta. Two complementary but differing QFT's explain with unrivaled precision what matter and light are, **Quantum Chromodynamics** (QCD) and **Quantum Electrodynamics** (QED) respectively.

QCD describes the core structure of atoms, going all the way down the rabbit hole to arrive at the kaleidoscopic world of their nucleons; the tempests known as protons and neutrons. These nuclei that make up the heart of matter form the lion's share of mass. QCD explains their mass-generating nature through a curious process Wilczek has dubbed "Mass Without Mass."[†] This is

---

[*] Davies, P., Gribbin, J.: *The Matter Myth*. Simon and Schuster. (2007)
[†] Wilczek, F.: *A Beautiful Question*. Penguin Books. (2016)

because mass, or inertia, known more intimately as solidity, is an emergent phenomenon, that arises from processes and dynamics that are not themselves massive.

In the space of a picometer, the core of an atom is an incredible maelstrom of energy oscillating near the speed of light. It is this confined torrent of power that allows nature to create solidity from ethereal fluidity — achieving Mass without Mass. QCD explains what all baryonic matter — that is, the substance that makes up the solid objects of the universe — is. It has it that the world is "made" of just three types of quanta that arise as vibrations in their fields. QED, on the other hand, homes in on the scission between Maxwell's angel (light) and matter, explains their true nature and how, through their "virtual" interactions, we arrive at electrochemistry and the table of elements.

Troubling for naïve theories of causation, quantum theory seems to demolish determinism and replaces it with inherent probabilistic randomness. In classical physics, solid objects (point particles) have well-defined trajectories — precise positions and velocities — such that one can predict or determine beforehand where a particle will end up given specific initial conditions. QFT's principal object is no particle but rather a quantum described mathematically by something called a **wavefunction**.

A quantum is itself a wave, a highly unified, spatially extended, specific quantity of field energy. **Quanta** (the plural of quantum) exhibit two particle-*like* epiphenomena and have the capacity to shapeshift into one another. That is, they are countable and localize upon interaction to a smaller volume, but not quite a point. As spatially extended energy ripples, a quantum's "position" and "velocity" possess an intrinsic and interrelated range of uncertainty.

This interconnected range of uncertainty is known as **Heisenberg's Indeterminacy Principle**. It's a consequence of quanta's waviness, and it bakes true randomness and uncertainty into the theory. Even if we knew absolutely everything about a system, we are no longer able to determine where a "particle" may be, but rather, we are left with only the capacity to calculate the probabilistic location where quantum waves — quanta more precisely — will interact.

The **Born Rule** is commonly considered to reveal the probable location of where we may find a particle should we choose to interact with a particular system. But in realistically interpreted quantum physics, the Born Rule need only tell us of a wave's *probable* interaction location; the 'where' a quantum wave is said to **collapse**. This **quantum state** or **wavefunction collapse** is a contentious issue, with most not agreeing on its reality as it implies a kind of instantaneous action that many theorists are uncomfortable with. But if we are to understand the theory, we must also accept what experiment is telling us and see that it's due to quanta's discrete and unified nature they are capable of collapse and that their interactions will *appear* to us as particles.

We can decompose the process of collapse into its three overarching elements. First, any interaction with a quantum system will be seen as *a discontinuous break* in the unitary time evolution of said system. Second, this discrete action *randomly "selects"* just one state out of the many it concurrently just realized. And third, that *the probability of this outcome* can be calculated by squaring the wavefunction to remove negative values. In a word, collapse implies discontinuity, randomness, and probability.

Admonishing the reality of collapse, the discrete, digital nature of Max Planck's original **quantum of action** hypothesis has instantaneity written all over it. Adding to this, **Bell's theorem** has been experimentally verified and *proves* that distantly entangled quanta adjust their states instantly and accordingly whenever a measurement is performed on either.

A crucial aspect of a realistic interpretation of quantum physics rests on accepting the collapse. But with it comes a better understanding of all other at first bizarre quantum behavior. As far as relativity is concerned, collapse does not violate it as there are no signals sent at the moment of it happening.

How do we know that they are delicate energy ripples characterized as waves, and given that, how is it that they gather to become solid localized macroscopic objects? An answer to the second question will be given later but to answer the first we need to know what a wave is. Put simply, a wave is a disturbance or dynamic shape of some medium. Quantum waves carry energy and information — as themselves — and are physical objects in their own right.

To see my meaning imagine setting up a wave in a rope. As you pull up on the rope (the medium), its **elasticity** will determine how much energy you must put into it to get it waving and how quickly or easily the movement will be transferred to the next section. Thus, the medium's elasticity directly affects the kinds of waves it can sustain and will require *a minimum amount* of effort for the wave to even set up. This "minimum amount" is our quantum of energy. Furthermore, the number of times you shake the rope will also be recorded, and the rope will carry that information along and transmit it to the wall or other person. And so, we see a wave — a unique thing to begin with as it's no "solid object," nor is it "local" to any one location — can be and thus possesses energy and information.

As we've seen, the media for quanta are vaporous kinds of elastic nothingness known as fields. Fields are entities in their own right; they exist and are no mere mathematical formality. Their ethereal nature allows them to possess novel characteristics and astounding capacities that lead to some truly bizarre behaviors, behaviors impossible for particles. For instance, as a spatially extended wave, already a "nonlocal" object, a quantum has a possible value — an "expectation" value — everywhere that it is. Owing to this, we can understand this "everywhere" as the fact that a wave dispersed throughout an environment, and with respect to the field, will have a probable value at its many occupied locations. That is to say, it is natural for nonlocal waves to embody **superposition**.

Strictly a wave phenomenon, the principle of superposition is key to understanding quantum theory. If we thought in particle terms, we might erroneously believe a particle to be in two (or more) places at once, which is absurd. But in realizing the nature of quanta as wavy field resonances with varying values at multiple locations, superposition becomes a pure intuition about a wave's unitary yet spread-out character.

If superposition weren't enough, quantum fields also allow for their quanta to *entangle,* leading to a more explicit nonlocal connection between two or more field quanta. The term **entanglement** denotes the process by which two or more quanta merge, exert forces on and exchange information with one another, only to depart ways and behave as if they had become One through the process. Two quanta can come together, interact, and in leaving the interaction not only carry off information about the other, but have become one another. The two will thereafter behave as a unit as if the two have become a single quantum and will thereby share a single quantum state.

There are many quantum fields, each harboring its own unique resonant quanta. When they are "activated" by their quanta, it will exhibit and appear as a certain effect. Broadly, we can categorize quantum fields into two overarching classes: matter fields whose quanta are classed as **fermions** and force fields whose quanta are classed as **bosons**.

Matter fields harbor quanta that "**spin**" in such a unique way that they acquire mass (inertia, resistance to acceleration), and despite being "matter waves," they nevertheless remain quite ethereal, *almost* insubstantial. These quanta achieve inertia by involving additional fields in their constitution, specifically: the **Higgs** field. The Higgs itself houses the famous "god-particle" — the Higgs boson — but, more comprehensively, the Higgs is an omnipresent scalar *field* whose energy content is everywhere non-zero. By involving this "force" field in their constitution, matter quanta feel and respond to the potential it creates, and the result is inertia. But this inertia is not yet mass as we know it.

Excuse the pre-emptive technical jargon, but to achieve "actual solidity," **color confinement** must beget **asymptotically free** quanta by necessitating that they operate in unison as a necessary relation, specifically, as the link between **quark** fields as matter fields and color **gluon** fields as "force" fields. It is only by the simultaneous involvement of both matter and force fields that the marble hearts of atoms — nucleons (more commonly known as protons and neutrons) — form. Quantum chromodynamics describes the dynamics and ontology of these marble-like tempests. Not to belay the point, but it is only by their distillation into actuality that the world of matter and mass can come into being.

Immateriality at the Core of Quantum Theory

As for "force" fields, the word itself is a bit of a misnomer as there are no real "forces" in quantum theory. Instead, what is seen by us as a force acting between two "material" objects — electrons say — is simply an exchange of energy quanta between them via an intermediary field. This field, in the case of electrons, is the electromagnetic. Stated differently, a force field is nothing but an "in-between" field whose function allows for, connects, and registers the transformations of the quanta of matter fields.

As for these "transformations," any matter field's quantum can randomly shift its **phase**. This "**phase shift**" is a moment when the shape or structure of its wavefunction undergoes an unalterable, spontaneous, and discrete *change*. For intuition, know that a wave's phase is captured and computed mathematically by modeling it as degrees around a circle. When a quantum shifts phase, it is as if a part of its phase — a "place" in its shape — suddenly jumps from, say, 10 to 20 degrees. The consequence of this "phase shift" is that the form of the very wave itself suddenly alters in an ugly, discontinuous way. But in a way that must be recognized and registered by distorting another field.

The only way that the quantum in question could accomplish this is if another entity could propagate the change away, i.e., an intermediary force, or better, interaction field. Stated in terms of quanta themselves, when an electron "phase shifts" this action is accounted for by the simultaneous creation of a disturbance, a "virtual" photon, in the EM field. My apologies if all this seems like much but quantum theory is vast, and I hope to accomplish here only an overview of some of its many core concepts.

♀

Both QFT and GR are "locally symmetric" field theories but at present remain disunited. Locally symmetric means that their transformations must be relayed from their point of origin through an intermediate interaction field more commonly known as a gauge field. The name "gauge field" is a historical accident as they are more accurately *phase fields* as they exist as a symmetrical consequence of altering the *phase* of the wavefunction.

For instance, in QCD, a "blue" quark may "turn" into a "red" quark and the transformative difference between the two is carried away as a "gauge" wave, in this case, a gluon. That "turn" is actually the quantum's phase shift and it coordinates the conservation of said quantum's **charge**, "color" charge in this case. Color charge "is really" a quarks capacity to produce and feel phase shifts, while electric charge "is really" an electron's capacity to produce and feel phase shifts. "Is really" because that's the deeper explanation for charge

69

but remains in scare quotes because there could exist an even deeper reason for it. Such as a loop of string... But we are getting ahead of ourselves...

There are two major crises in fundamental physics: the disunity between QFT and GR and interpretational problems regarding the foundations of QFT itself.

Known as the **background (in)dependency problem,** the crux of the incompatibility between QFT and GR lies in how each theory treats spacetime. The reason being that QFT operates on the 'flat' spacetime defined by Special Relativity (SR), while GR, utilizing a geometrodynamic approach, describes how energy-momentum warps spacetime such that its resulting curvature realizes itself as gravity.

As for QFT, its arena is known as "Minkowski spacetime," as it was initially conceived of by Hermann Minkowski, an old, and once-skeptical, professor of Einstein. After having read Einstein's papers on SR, Minkowski went on to give it a sound geometro-mathmatical basis, with the result being the very *unification of space and time.* Initially, Einstein disapproved of Minkowski's overly mathematical treatment of SR, but later recognized its importance as it came to serve as a fundamental necessity in his quest to "generalize" relativity to include gravity. Unfortunately, Minkowski did not witness the fruition of Einstein's generalization of relativity as he passed away in 1909 due to appendicitis.

As for the nature of Minkowski spacetime, it is a 4-dimensional manifold comprised of three dimensions of Euclidean space and a single dimension of time. A flat, static, isotropic, and infinite scaffold where "events" share a common "spacetime interval" irrespective of inertial frames. This pre-GR spacetime serves as the stage upon which QFT unfolds and its actors are made to dance. It acts as a *fixed* background structure and remains *unaffected* by the quanta present upon it. Consequently, the defining equations of QFT remain *independent* of the specific shape of this spacetime, making QFT *background independent.*

On the other hand, GR is inherently *background-dependent* because its equations *depend* on the shape of spacetime. Massive objects are but large conglomerations of energy-momentum that *affect* spacetime itself, their stage-like scaffolding. Again, GR describes how large densities of energy-momentum cause the curvature and warping of spacetime, the result of which is gravity. Unlike the flat spacetime of QFT, GR's spacetime becomes warped by the massive objects present to it and this "curvature" is understood as gravity.

Obviously, this issue of background (in)dependency constitutes a significant barrier to their unification. It is possible, however, that it may only be solvable in light of philosophical, rather than physical, arguments.

That is to say, maybe a complete overhaul of our presumptions about the true nature of reality is necessary. Maybe "Total-Being" is not simply a material-energetic cosmos obeying physical coherency laws but is instead an unfathomable Infinite whose possible modes of realization result in an ontologically pluralistic palace.

Or maybe it is the case that the mereological modalities of the cosmos differ in form and function (something similar to the relations between fields of sense) when made to move between certain levels of abstraction, scale, and magnitude. Maybe the quantum nature of the lower-lying fields simply becomes irrelevant when we level-up to the larger-order domain of distinction that considers gravity? Anyways, I digress... As this could become a long exegesis, I should only like to point out that I do believe philosophical arguments are more relevant then ever for this discussion. Physicists may not be happy, much less satisfied, with this suggestion, but it still may be the case that novel forms of philosophical reflection will furnish us with a deeper understanding of the issue.

Now, of the crises in QFT alone, we saw earlier that it is the formalism's silence on a physical ontology — precipitated by the undecidability among professionals as to the ontological meaning regarding the 'what' of the wavefunction — that forms its core issue. The theory doesn't say whether matter made of fields, waves, or particles, except that "energy," whatever that may ultimately be, is its most essential "substance."

Why we don't know for certain and are instead forced to *interpret* QFT is due to the famous "measurement problem." This "measurement problem," is in fact, two interrelated problems called **The Problem of Definite Outcomes (PDO)** — more commonly known as Schrödinger's cat — and **The Problem of Irreversibility (PI)**. Their suggested resolutions lead to quantum theory's many interpretations.

The PDO stems from the fact that the theory says that quantum systems evolve into a superposition of states, rather than the single definite outcomes that experiments observe. The PDO analyzes how — through quantum state collapse — an indefinite state of superposed potentialities always results in a single definite outcome. Schrödinger himself — whose equation defines the evolution of quantum systems — believed it to predict a rather "ridiculous case" of a cat that is both alive and dead at the same time.

The PI stems from the fact that the unitary time-evolution of Schrödinger's equation is reversible, a seeming violation of the second law of thermodynamics. But what if both problems could be solved by adhering to a realistic view of QFT? A view that accepts the "at-first-glance" counterintuitive dynamics of quantum fields; collapse, superposition, nonlocality, decoherence, and entanglement. All of which are easily intuited and understood when one drops "particle" from their conceptualization scheme to realize that waves easily display these capacities and one can solve both problems without appealing to any kind of quantum mysticism.

As mentioned previously, QFT is interchangeably referred to as **The Standard Model of Particle Physics**, leading many to consider it a theory of the tiny and too small to see; subatomic particles and the fields of forces that move them about. But as previously asserted, QFT is a theory of the very fabric of the universe, and its objects are not even necessarily small or particulate. In the future, as the true image of QFT is revealed to the public, the names "the standard model of particle physics" and "quantum mechanics" will fall to the wayside. QFT is certainly not about tiny solid objects or particles and is therefore not a 'mechanics.'

I am here taking a cue from Art Hobson, who — if there are not Many-Worlds — has achieved a timeless creation with his book *Tales of the Quantum*. This work will go on to become the go-to manual for realistically interpreted quantum physics. A particularly salient reason is his commitment to language. He consistently refers to quanta as quanta while also dispensing with talk of 'mechanics' and 'particles' and instead speaks plainly of waves in fields and their capabilities and features such as nonlocality, superposition, and entanglement. Furthermore, he has rediscovered a simple resolution to the PDO, originally put forward by Joseph Jauch, showing how Schrödinger's original analysis was flawed and that it can be resolved by analysing the zombie cat's measurement state and taking seriously the nonlocal nature of decoherence effects. Hobson calls this the "local-state solution" to the PDO. A solution we will later examine in detail.

☿

Now, one might wonder about what all quantum fields have in common, aside from being fields. The answer is that they all require a communal and shared "carrier." This is commonly known as the "vacuum," but the concept of which does little to illuminate just what exactly it itself is.

Doing a little better as a stand-in, I aim to resurrect a concept with an ancient history, that of the 'luminiferous' Aether. Harking back to Aristotle, this "quintessence" as primal substance turned 'vortex sponge' by Maxwell, was once thought to be the mechanical medium by which his angel (light) took flight. Historically, we are told the experiments of Michelson and Morley, and later, Einstein's theory of relativity, dispensed with the notion of an Aether. However, they only showed that a concept of the Aether as a mechanical medium or fluid at absolute rest is impossible. A result that does not negate

Immateriality at the Core of Quantum Theory

the possibility of a refined Aether, one with different attributes, capacities, and properties.

Again, the aether concept disproved by Michelson and Morley's experiment was simply a possible "frame of absolute rest." A frame Michelson and Morley thought they could measure themselves against. But no such frame of absolute rest exists, and motion is truly *relative* as it is only by comparing between frames can one deduce motion. Nevertheless, an aether thought of substantially, even if massless and immaterial, as the carrier of fields and their waves must exist.

Although part of me wishes to reserve the name 'Aether' for the four-vector EM field that sustains and allows for the propagation of Maxwell's angel as the name is appropriately synonymous with the ancient Greek god of light. However, I will use the term Aether to refer to the quintessence of Aristotle, the "fifth" element thought to interpenetrate all including the heavens above while restricting its 'luminiferous' aspect to that of the EM field. Put bluntly, the theory of relativity, geometrical requirements and local symmetry constraints *require* the existence of an Aether as a common carrier whose most "basic" field may be the metric manifold of spacetime itself.

Again, we ask, just "what" is "the" vacuum? Humble physicists will be the first to admit, they do not truly know what the vacuum is; just that it serves as the carrier for various fields and possesses certain bare qualities that we can measure and ascribe to it, but the 'what' of it remains a mystery. Regardless, it's the very ground out of which all things come, so how can we familiarize ourselves with it aside from ingesting psychedelics or sitting in transcendental meditation? Well, we could create one...

Imagine removing all the dust and air from a glass chamber such that we remove every atom from a region of space. We have created a vacuum, but what is left? Somewhat paradoxically and naively, we know nothing is there because we see right through the glass chamber. Yet there is something there, you just cannot see it even though it's what sustains the light that reaches our eyes. It's the EM field that permeates the whole of the universe and one cannot remove it from the chamber for it's a characteristic of the vacuum itself.

The EM field is a vaporous "layer" of the vacuum – a distinct characteristic – it is the luminiferous aspect of the Aether. So, our chamber is not truly empty, a field is there and the only 'thing' a field can sustain is a dynamic shape of itself, most commonly, a wave. Is there anything else in there? Yes, a gravitational field also permeates the chamber and the room equivocally. The fields go right through everything, but their quanta do not. The quanta of fields are forced to interact through the potentials they "feel." In the chamber would also reside inactive or "lowest energy state" quark and gluon fields, a neutrino field, **Higgs field**, electric field, and so on.

Immateriality at the Core of Quantum Theory

In Aristotelian terminology, we could equate the vacuum-as-carrier as a "substance" that stands in contradistinction to its "attributes," which would be its various fields. That immaterial "substance," however, is 'like' nothing; a "nothing" that is an ocean of existential potential. To intuit its "bare" existence, consider the Casimir effect. Imagine two ships afloat in a calm sea a meter apart. With nothing happening, nothing happens. Fascinating! Now, disturb the sea and ask what will happen to the ships. Will they float away from each other or come together? The answer is to be found by considering the various wavelengths that would set up upon disturbing the sea.

Consider how many different wavelengths of waves can exist and persist in the space *outside* of the ships, whereas only a few limited wavelengths can fit in-between the ships. Given that, one could expect a net force on the ships that will push them together and that's precisely what happens. The same is true of the vacuum as this 'Casimir' effect happens to uncharged metal plates placed in the abyss of empty space. They are "drawn" together because there exist more options for virtual quantum fluctuations of varying wavelengths to come into being *outside* of the plates than there are in-between them. If virtual "particles" were popping in and out of existence around the plates they would not "attract," as they would form a net equal effect. As such, wavelengths — even virtual ones — are key to understanding the Casimir effect.

That "nothing" inside that chamber is what this entire book is about. It is the baseless fabric of this vision, and it harbors "consciousness" as yet another field. It's the very ground for everything you see, feel and hear all around you. The "vacuum" as Aether is the carrier of the invisible fields that interpenetrate the chamber, the room, and the whole universe. These fields are best intuited as properties of space, but they are more accurately *bare, structural qualities* of the void-vacuum. A "dressed" electron is more than a quantum of energy oscillating in the electric field, it must also include in its constitution and thereby direct some of the energy that it is into deforming the EM and Higgs fields.

As such, for an electron to be a thing involves a multiplicity of fields, and having achieved a novel unity in itself, becomes a quantum belonging to its own field: the electric. In a similar way, the field of consciousness (or mindfield) possesses a similar constitution, involving many in its eventual realization. And despite its advent as temporally isolatable, whose realization or activation is to be understood as the genesis of life, consciousness can nevertheless be seen as an always-already "there" phenomenon, thus sidestepping the **cosmic late-comer problem**.

To intuit the Aether as a kind of carrier *preexisting* the dawn of Time, remove all fields and try to grasp what would be left; is it "nothing" or the indivisible Absolute? All we can fathom is an infinite abyss, but rather than a void, we know it as a Plenum. There is wholeness, fullness, and dynamism in

Immateriality at the Core of Quantum Theory

this darkness. In this virtual sea, pre-Big Bang purgatory, we are reminded that the Casimir effect tells us that this oceanic darkness never remains still; like wind causing ripples on a pond, virtual quanta endlessly oscillate. Is it possible that in this eternal emptiness, somehow, virtual quanta superposed in such a way as to become a "rogue wave" and cause the Big Bang? Tolling the bell, a cacophonic symphony of increasingly complex waves set the void vibrating, becoming "energy" that it itself is. Is this how the Universe started?

Could "the vacuum" also harbor within it the propensity to activate itself as a universal field of consciousness? Consciousness is an embedded frequency analyzer that decodes and specifies a localized and temporally extended portion of this vibrating void/plenum. We know by living it that consciousness is physically integrated and causally active in the universe. At least as physical as that no-thing that is a thing (a field) carried by the vacuum. Moving the other way, we could invert the argument and claim that what we commonly conceive of as physical is as vaporous and ethereal as the very mind that contemplates it. Regardless, the essential notion and claim of this project, is that the two are of one and the same ground. We are all borne of a unified and common carrier. We are a void set vibrating; an abyss eternally resonating. The music of spheres pulsating at frequencies of unimaginable scale — entangling bits and superposing waves — all to become *us* and the furniture of reality.

What we know and live as consciousness is a feature of the vacuum that we — as historically differentiated conscious agents — *access*, and the degree to which we do is associated with the complexity of the information transacting organisms that we are. Our visual field is our consciousness extended beyond ourselves because the contents of our consciousness is ourselves. When we share space as agents, the wavefunctions of our consciousness overlap or superpose such that we both are able to, but not necessarily will, *know* one and the same thing; a beautiful scene say.

Truly, we do not know what fields are, save for we find equations that agree with experiment that corroborate their definition as properties of space. They tell us the thing that moves through them is a kind of wave. A complex or standing wave that becomes a well of multi-directional influence in and of the field.

Structuring "Layers" of the Vacuum

Beneficial to our stretching imaginations, we can return to and utilize the background dependency problem to conceptually stratify the vacuum and consider how one might build up larger-order structures from lower-order processes. At "bottom," there exists a unified, immaterial, prespacetime vacuum, carrier of fields and the mathematical, semantic-information-theoretic, complex-algorithmic, computational capacity and machine-code of the universe. This "layer" is dynamic and a quantum-information interpretation of QFT places this "layer" on a pedestal. This underlying "sub-quantum" field contains what David Bohm calls the *quantum potential* field, as its objects are not yet entirely actual, and it harbors an implicate or enfolded order of information. That is, an invisible order encompassing discrete elements.

The theory explaining the first layer of the explicate order — the world of form made manifest "atop" of the implicate informational layer — would be general relativity and energy-momentum's effect on spacetime. Although spacetime curvature may have a deeper origin in the implicate layer as it may be that entangled information is what impacts the universe's geometry. As we saw before, it may be that space itself arises from the relational information shared between entangled qubits and that it is this process that allows the universe to define distance—the 'how' it is able to set a metric. Intraatomic structure as revealed by QCD would then sit "on top" of that. In the calculations of QCD and QED, gravity is negligible, allowing the theory to effectively ignore the curvature of spacetime.

Let us continue; "above" QCD would lie the realm of QED and the electric-elemental world of chemistry. When electro-molecular interactions and the complexity of their entanglement reach a certain level, they open the doors of perception and the domain of biological processes. How this comes about — the abiogenesis of Life and consequential consciousness — remains an enigma.

How exactly does seemingly lifeless electrochemical energy reach a certain threshold of molecular complexity and "come alive?" This is a grand question. As we have seen, this book's thesis intends to give at least some understanding of that question by arguing that "mind" is always-already "in" the implicate order awaiting "activation." The underlying complexity of local information processing by entangled quanta opens the door to mindedness, and as its algorithm *evolves*, as so too does the depth of awareness penetration. That is, the deeper you go, the greater the amplitude deviation of the mindfield, the more "expansive" the felt intimacy of consciousness.

As we know, biology then opens the door to psychology, sociology, and mindedness in general. Looking back, we observe that the appearance of the rational mind into the historical record results in an explosion of ontology; where ideas come into being and begin to affect existence. The human brain is an ontology engine that, by treating reality as clay, creates artefacts like computers and keyboards.

If this layered stratification is correct, certain functions build up higher features, which, in turn, "supervene" on lower features, creating a feedback

Immateriality at the Core of Quantum Theory

loop that is more reminiscent of a strange loop's involution of itself. A two-way street whose directions indicate reductionism and emergence. But as we shall see, QFT supplies a fascinating mereological model that transcends the operant paradigms of reductionism and emergence. This is because its unified systems, always remain, at their cores, collective distortions of their underlying fields. And although this layered stratification may be conceptually useful, it is superficial because we want hold to the notion of the unity of reality and see that these layers are One.

Each of the "certainly" physical orders previously considered — the information-theoretic, GR, and QFT — form a different "layer" present in the (possibly prespacetime) vacuum. It is useful to point out; all these theories are physical, though not material, and are present in or upon "the vacuum."

$$\Psi$$

We have said much of the fundamental-analytic portion of Being (its essence extended in Space) but what of its Doing (its temporal essence expressed as Time).

Of the complex-algorithmic, recent developments in quantum computation and computer science have ushered in a new participatory ontology based on the computational processing of information. As Noam Chomsky is said to have mentioned in conversation,[*] the moment we get a clear grasp of something, we deem it physical. Indeed, one hears the common adage today that "information is physical." Implied in this sentence is a more profound implication that what we mean when we say something "is physical" is that it's *real*. It has become a quantifiable quantity that really exists. Consciousness is also real, and an innovative approach to it — **integrated-information theory** — is attempting to quantify it.

Thus, the superfluous entity known as information lies "underneath" QFT and GR. As such, they are underscored by a common-to-all Information-Participatory (IP) ontology, and it is this framework that just might bring about their unification. The "participatory" part is indicative of the fact that every quantum interaction is about the *mutual* exchange of information, thus, any and every interacting system, participates in its own constitution via its contact with others.

As MIT Professor and quantum computer scientist Seth Lloyd wrote: "The universe is *indistinguishable* from a quantum computer" because "quantum computers process the information stored on individual atoms,

---

[*] Searle, J, R.: *The Rediscovery of Mind*. MIT Press. (1992)

electrons, and photons." Thus, this underlying, information-mathematic implicate order is physically real but causally inert, requiring energy for its explicate expression. Without energy, we would never arrive at an "it from bit." Also, without the addition of energy injected into the universe, quantum fields, too, would be causally inert.

As information processing possesses an inherent element of dynamism, it forms the backbone of the complex-algorithmic domain of causation, the bookkeeping ledger of the machine code's universal unfolding. Furthermore, information and entropy have also been shown to share a deep link and can even be stated in terms of the same equation. Entropy is a measure of the invisible information contained in a system. Furthermore, in the quantum context, qubits of information can entangle such that their relations — their now-coupled degrees of freedom — may come to define a metric, becoming space as we know it.

Indeed, were it not for information entanglement, achieved by the mutual contact of objects in motion, space (and possibly time) would be without sense and meaning. Consider, the very idea of space only begins to make intuitive sense when we observe how two material objects stand in relation. Astonishingly, it may be the case that even for something so fundamental as space to exist, at least two objects — be they qubits or whatever — must stand in participatory relation. Another astounding claim has it that this mechanism may also allow us to reconsider the Big Bang as a *creatio ex nihilo*, that is, how the universe moved the needle from zero to one and got something from nothing.

Suffice it to say, there would be no world, nothing to compute, if it were not for energy. Given this, any physical ontology must recognize that we require, at minimum, the trinity of information, fields, and energy to create a "substantial" world. Is it not amazing that of all that has been written about thus far, not a single thing has been fundamentally material? We have not yet arrived at inertia or mass.

Finally, to accentuate this dynamic, complex-algorithmic mode of Becoming, David Bohm — echoing Heraclitus and Henri Bergson — reminds us that: "*What is* is the process of becoming itself, while all objects, events, entities, conditions, structures, etc., are forms that can be abstracted from this process."

☉

What has been revealed by our brief foray into QFT and GR is that the interwoven fabric of reality is a "carrier" of waves of information-carrying energy fluctuating in unbounded fields. Sometimes even just vibrations of spacetime itself as in the case of recently confirmed gravitational waves. To wit, the fundamental fabric of the universe is a set of interacting fields underscored by a common-to-all geometric-dynamical spacetime structure, or metric field, whose energetic interactions register as the computation of information.

Given the previous discussion, concerning the physical ontology of the universe, one can consider one of seven "entities" to be the "most" fundamental, none of which are material and only 'energy' should be regarded as "substantial." And although one cannot *reduce* everything we encounter to any of these or their interplay, we will gain significant insight if we analyze them. These are the vacuum, spacetime, fields, energy, spin, information, and consciousness. However, we could reduce that number to six by arguing that spacetime is but another field.

One of these may be the most basic "substrate" of reality. In my view, and if I were to stratify it based on what I believe to be the order of fundamentality, I would list it; consciousness = prespacetime vacuum, spin, information, spacetime, fields, and energy. Setting aside ruminations on consciousness and the fact that it's as physical as the other entities just listed, we will proceed with the "certainly" physical. So far as I'm concerned, consciousness is at least as physical as information, and certainly as tangible as the EM field.

Consider, the EM field can burn your skin, but another's Will can break your bones. The "force" of consciousness is known by it as Will. Also, keep in mind, as will be explored in more detail later, that reductive materialism imploded on itself, revealing immateriality to be at the core of things, placing mind and matter on equal footing, and opening the door for mental causation and mental realism. At the same time, the random and probabilistic nature of QFT gives weight to the freedom of human agency.

The future is not yet decided, but is, in every instant, constantly being calculated, as the quantum system that is the universe continuously and incomprehensibly evolves and collapses, over and over again. In another work, I refer to this structure of quantum evolution and information-exchange collapse as *the pulse of Being*. Within it, we are liberated with a physically real argument for the freedom of the Will. Finally, although we consider consciousness as physical — possessing the capacity to compute information and affect the furniture of the world — it's something somehow *underneath*, yet *above and beyond* it. Even if consciousness comes about through information processing and energy's evolution and therefore has some physical antecedents and prerequisites, it is nevertheless different in kind and quality. In a word, consciousness is in a class of its own.

Finally, let us turn now to get an initial grip on fields, their structural types, and attempt to analyze "the void" they may overlay — and the nature of reality as immaterial in general — let's turn now to some ancient myths and metaphorically associate them with the prespacetime vacuum.

Immateriality at the Core of Quantum Theory

Another misconception is that the quantum, being discrete, somehow pixelates the fabric spacetime. Quanta are smoothly varying waves that *because* they are waves, oscillate in discretely different "modes." Furthermore, only their energies are quantized and the qubits of information they contain are binary. It is for these reasons — a digital (as opposed to analog) energy value and the dichotomous nature of information — that serves as the quantum's unifying function. Some might say that spin is quantized and that is correct, when measured it will always be found to be this or that.

> "There are no differences but differences of degree between
> different degrees of difference and no difference."
> – William James, under nitrous oxide, 1882

CHAPTER VII

Mythologies of the Quantum Foam

In Norse mythology, we encounter the concept of Ginnungagap, a "magical and creative power-filled space,"* an eternal and ethereal void with a "magical charge" that predates creation and out of which a strange cosmogony is born. It is not the resulting cosmogony that is of interest to us but the early intuition of a 'void with a charge' because an especially important quantum field is somewhat like that. Also, the name Ginnungagap helps to hide our ignorance of the fact that we don't know what existed pre-Big Bang or what the vacuum actually is. We are, of course, referring to the famous Higgs field — the field responsible for engendering mass — is like Ginnungagap: it is a "nothing" with a "magical" charge. Its resonant quantum — basic unit of waving energy — is the famous Higgs Boson.

The Higgs can be considered a property or function of space itself (the void) that possesses a fluctuating, always positive, non-zero energy throughout — it is "power-filled." Its "almost" a medium or substance, but is more accurately a scalar, universal field that permeates the entire universe. Scalar means that it is a kind of "stationary" fluid that, to represent its energetic magnitude, requires only a single number at every point in space. That is because the Higgs isn't moving; it is still waters.

When this omnipresent, fluid-like energy-field couples to other fields, it creates a potential that provides a mechanism by which wavy quanta take on the property of inertia, thus acquiring mass. That is, provided the adjacent, spatially inseparable, coupling field possesses the appropriate feature to do so; namely, an internal degree of freedom called half-integer spin. Surprisingly, this amounts to only one type of quanta; fermions, which themselves fall into the two classes that make up the quanta of matter-fields — electrons and quarks. Later, we will have a lot more to say of Fermions, Bosons, and spin.

---

* de Vries, Jan.: *Old Norse Etymological Dictionary.* Leiden: Brill. (1977)

There are multiple quantum fields that can "layer" to form more exotic fields. Consider the strong interaction, quarks (themselves fermions) and gluons (bosons) gather to form the nucleic hearts of atoms — protons and neutrons. These, surprisingly, a "level up," are again, themselves fermions. Each field has its own resident, resonant quantum. For the EM field, it's the photon; the electric, the electron; the metric, the graviton and so on. Foregoing the fact that quantum fields are "operator-valued," while accentuating that they bear as themselves a determinable property, a fields configuration is designated by, and restricted to, its structural type: there exist scalar fields like the Higgs, as well as vector, tensor, and spinor fields.

A scalar fields configuration requires only a magnitude at every point. Take the atmosphere as an example of a carrier of fields. The air in the form of a gas is a medium. This medium can have a "temperature field," which at each point in the medium has a thermodynamic value or magnitude. It may also have a moisture density that varies from place to place, a kind of "moisture field." All these "fields" can only exist by virtue of that which allows them to be, their "carrier," the atmosphere.

Distinct from scalar fields are vector fields. Vector fields are a kind of dynamic field that possesses both a magnitude *and* a direction. To represent them, a number for each is needed. If one considers the atmosphere as a field of air only, it is a carrier. If one adds a pressure or temperature grid to the air, we can see each as a distinct scaler field. Finally, if we add swirling winds and measure their speeds and directions with weathervanes and anemometers, we uncover a vector field and find vector quantities all over the map. Having added direction and magnitude to the "air" gives us a good intuition for seeing fields as a kind of fluid. Electric and magnetic fields are vector fields, and so too is the field of their unification, the EM field.

Like the Higgs, this field permeates the entire universe, yet its bubbling activity — its quanta, photons — move and have a direction. Light — in the form of coproducing electric and magnetic fields — propagate through the EM field as a complex, spherical wave with a well-known velocity. It is complex in both its mathematical and naïve connotations. Thus, photons — quanta of light — are energy-ripples in a massless vector field. Although photons are "polarized," they do not "spin" in the appropriate fashion and, as such, do not interact with the Higgs.

With respect to complicated degrees of freedom, the most complex types of fields are tensor and spinor fields. At every point in space, a tensor field is represented by a whole grid of numbers — a matrix — that tell how vectors distort and warp in that location. Distinct from a vector field as having only a single direction, the tensor field is omnidirectional with varying degrees of magnitude.

Tensor fields are present in solids; think of what happens when one twists or bends Jell-O. At every point, as the initial twist propagates through the gelatinous medium, the vectors change in magnitude and direction and relay that on to the next point. The universal field of gravity, the metric field, is just such a tensor field where localized aggregates of energy-momentum (matter) distort the immaterial "Jell-O" that is spacetime itself. She represents a constant sink wherein matter attracts itself due to its warping effect on spacetime. Curved, indiscriminate, and underlying all fields, she unceasingly pulls everything into her bosom.

Finally, a spinor field is equivalent in complexity to a tensor field but quanta in a spinor field "spin," a process we will later consider in detail. Spinor fields describe fermions.

Fields are not themselves energy. They are properties of space that allow for and regulate the fluid movements *of* energy. Energy is necessary for displacing the field at a location but is and is not itself the field. The two are distinct but intimately related. In fact, for some fields to be effective, they must come together. Spacetime, however, is somewhat different. Gravity is not energy but the "force" of the metric field, geometrically understood as the "curvature" of spacetime caused by the presence of localized energy-momentum. Gravity, like light, is a field and is itself massless and immaterial. There is a distinction between energy and fields, but as we can see, they are inexorably linked.

Not to belay the point but QFT and GR tell us that present in the vacuum lies a multiplicity of fields, each interwoven within one another, "occupying" the same space for they extend throughout all space at all times. The Democritian view of a void is denied. We live a fullness, a plenum; "empty" space does not exist. Reality is full, yet this fullness is hidden, veiled by its thin vaporousness; immaterial, it is "like" nothing. As such, the void of the ancients no longer a void but a misty incorporeal quintessence. Something is there where we always thought of it as nothing. As regards the metric field, physical research has resurrected the mythological titan Khaos, the mother of the universe and primordial abyss has returned and has given birth to her airs.

Humanities Core Theory stands as our most complete description of the natural order, and the concepts and ideas it contains possess an undeniable beauty. Nevertheless, the Core Theory is incomplete. It fails to address the problem of consciousness and leaves us "in the dark" concerning dark energy and dark matter. The word 'dark' in these names illustrative of our ignorance, not their properties. We've observed their effects and thus inferred their existence, but they remain a mystery. Could it be that both phenomena are fields whose effects are so large we are unable to test them? For as we have seen, the four fundamental forces explained by GR and QFT are all field theories governed by local symmetries. So far, we have been unable to reconcile quantum theory and GR; however, these theories share a typical structure, and so I don't believe it's a stretch to say they share a common ground or carrier. That ground, of course, being the quantum foam.

The Womb of Khaos – Newton's Cradle & Einstein's Pearl

In ancient Greek mythology, we encounter the Titan Khaos—the motherwomb of the world out of which all creation is said to come. She is the precursor to all things and gives life to all lesser gods—some so vaporous that they are personified as "airs." She underlies all and is the "mother of airs," a concept we can metaphorically compare to the metric field of general relativity (GR) as it "underlies" all and is the "mother of *fields*."

Khaos — the chasm of air — is spacetime itself. Womblike, her metric fluid accounts for and allows for the geometrodynamical transformations of spacetime to take place. Gravity waves travel across the universe at the speed of light, bending the ocean of space and time on their never-ending journey. Black holes are her pearls, magnificent orbs of onyx, wells of incessant wanting, pulling all energy into her bottomless bosom. They are Erebus, her first-born, God of Darkness and netherworld shadow, black holes are perfect objects capable of consuming her third born, Aether—God of Light and ethereal mists of the heavens. Woven throughout her fluid-fabric are her many "airs," her many fields. Ubiquitous and omnipresent, saturating the world with their reach, fields are everywhere and everywhen, modalities of an abyss, properties of a not-yet-a-thing.

GR describes the spacetime structure underlying all fields as a fluid-like *metric* field, a tensor field modeled as an absolute yet malleable 4-dimensional fabric. Spacetime entertains determinate 'events,' moments that are forever etched into the fabric of reality. These fixed events mark the *interactions* of, and between, quanta which are not in themselves determinate objects. A coherent, freely propagating quantum is determined — both in theory and practice — as *undetermined*. In other words, quanta evolve into nonlocal superposition states and as such become *definitely indefinite*. They come to be a way, many ways in fact, but until they interact, we know not how they are and can only calculate the probable locations where they *might* collapse.

As an aside, despite the common adage as to the supposed non-relation of quantum mechanics and human mindedness, that the nature of reality is of a quantum nature nevertheless suggests a practical model for the freedom of the will. The reality of the quantum as an "determinable indeterminate" reveals that parts of the universe exist in states of simultaneous actuality and

Mythologies of the Quantum Foam

potentiality. They are determined to be what they are as captured by their wavefunctions and historical genesis. Yet, their ability to superpose allows them to take *all* paths as they evolve through space and time and remain indeterminate. This malleability of the quantum allows them to transcend some of the 'order of events' defined by the underlying spacetime for in between them, it remains an actually existing possibility.

This, I believe, is how we ought to see the human agent, their consciousness simultaneously containing a degree of what is actual and what is possible, given what is given.

Maxwell's momentous insight into the true nature of light as the EM field bestows to spacetime and ethereal quality. An Aether that allows for the flight of his luminous angel as transverse perturbations of dynamically dancing co-creating electric and magnetic fields, two "properties of nothingness." What is fascinating about the early intuition that guided people's mythmaking was that both Darkness and Light were understood all-pervasive mists or airs. Today, we know the phenomenon of light and its counterpart darkness as mere states of the EM field, and so, in a sense, as an all-pervasive air.

The experimental quantization of the mother of all fields — the metric field — has not yet been verified but remains a field. We are still treading the long and arduous road to quantum gravity. However, new insights on the horizon concerning information and entanglement seem to be pointing towards finding a way to have GR emerge from QFT rather than "quantizing" a classical field theory. Furthermore, as will be seen in the future of physics, both the dark forces present in the universe are "larger-order" fields, or so I predict. That is, dark energy and dark matter are neither material nor energetic but rather properties of cosmologically large space that manifest themselves as physical effects. Dark energy — as verified by our telescopes and calculations — gives us a profound and straightforward intuition as to what we mean by a field. That is space doing something and, in the case of dark energy, expanding it. Of dark matter, curving it into a sink.

Furthermore, if we considered dark energy to be authentic, the pillar that is the conservation of energy would crumble as from where this energy "comes from" is unknown. Thus, dark energy is not energy per se but is instead a field. Likewise, the immaterial material of dark matter is also a kind of field effect. The christening of these fields as "dark" is yet another clue to their true field nature. Invisible, the "airs" of Khaos are "things" that are not themselves things. They are never seen but only revealed and understood by proxy. That is, our measurements, data, and coherency demanding minds require their actuality.

Immaterial energy disturbances in fields build the whole of the explicate order. This understanding of "matter" as an emergent property of the interactions of fields emancipates man's cognitive faculties allowing us to begin to bridge the gap between the worlds of the sensible and the intuited—the physical and the mental. It's no longer a mystery how the mind can move matter about as they share a common immaterial ground. However, the view of the world as built upon that which dreams are made may be difficult to comprehend. So, let's go back to discover how we came to know of the nature of fields and how light is a wave in one.

Again, what follows is a realistic account of reality. That is, what we are ruminating about what exists as it does apart from us. Even though our shared and personal thoughts about electrons are not themselves electrons, for they are indeed informed thoughts, they nevertheless speak about real electrons. Electrons obtain. Or thoughts refer. That seemingly innocuous sentence hides a deep philosophical monster; correlationism. The notion that thought and being are always interrelated such that we never access the world as it's in itself but only our cognitive relation to it. This apparent "transparent cage" we're epistemically locked in, unable to claw our way out. Knowing only our knowledge and not the world. Supposedly, collectively constructing a sociological hallucination, dreaming while awake.

A further metaphysical commitment is to the realism of our subjectivity. Moral thoughts and emotions exist and are real. As real as the sun albeit of a different sort. A difference in kind but based upon the same ground. A ground which is omnipresent and immaterial, of a field of unified yet differentiated consciousness.

For instance, once one has understood evolution, the entire web of life crawling on our Garden of Eden Earth becomes family. Of course, theories are no mere hypothesis. Derived from empirical observation and data gathering; we have teased truth from Nature. Although the Sceptic might charge that they are based on assumptions, they are nevertheless assumptions gathered from our inextricably linked and dynamic dance with nature. We reproduce our results and humbly observe that our understanding holds, well, for now. Yet, we know it's more than that. The Core Theory of physics lies on a well-established and tested foundation, a base that will forever remain, until the end of time, true.

"Life, that means for us to transform constantly into light and flame all that we are and meet with; we cannot possibly do otherwise." - Friedrich Nietzsche

CHAPTER VIII

Maxwell's Angel | Aether: God of Light

Quantum field theory tells us what reality is at its most fundamental and substantial level. Through it, we know that part of the interwoven fabric of the universe — the very ground of reality — is invisible, space-filling fields. These eternal 'airs' are universal property spaces of varying complexities, all superimposed within and interpenetrating one another. Some interact with one another while others are inert to the presence of their brethren. The 'vacuum' is their carrier, and their vibrational modes are the energetic interplay that gives rise to and forms all natural, physical, and naively "material" phenomena.

All fields occupy all of space, but not all field effects have a universal range. The fundamental mechanism holding the nucleons of atoms together — a mechanism called confinement — is just one such process. Confinement is a local field interaction, a property we will come to discover when we visit the theory at the heart of matter, quantum chromodynamics.

Nevertheless, fields occupy all space because they are, in a sense, space itself. Fields are continuous, open-ended, and are either infinite or grow along with the expansion of the universe. They may have even existed before the introduction of energy into the cosmic show. Fields — being properties of space — may even precede the genesis of Ginnungagap and outlive Ragnarök - The Fall of the Gods and the End of Time.

Although one can only make guesses as to those notions, we may ponder — if infinite — fields may even ground multiple universes next door. That is if the **cosmological multiverse hypothesis** is correct, and we are but one bubble universe in an effervescent and unimaginable Unbound. If timeless, fields may be the essential part of all worlds ever. We have already speculated about the beginning of time, but what of its end?

Imagine what might remain at the end of time; when all the stars have burnt out, and every atom's nucleons have decayed into electromagnetic dust. The only events taking place over unfathomable stretches of time are the silent, spacetime-tearing mergers of black holes. Due to Hawking radiation, in this once-again abysmal darkness, and over unspeakable eons of time, these obsidian pearls of fluid crushing spacetime begin to evaporate. Eventually, they will disappear, and the only objects left in the entire universe will be their remnant radiated photons: low-energy EM field vibrations.

From this point, it would take these final photons roughly a quintillion years to die out. And when they finally do, with no relation or reference, time and space lose all meaning, and no-thing, once again, exists.

Have we returned to the primordial state of Khaos? Are we back to a sea of potential, the quantum foam? Who knows, but let's use this conceptual backdrop to ponder some other equally unanswerable questions. If the photon has evaporated, is the field still there? Into what did the last photons and black holes dissolve? With no objects, no disturbances left, and space and time without sense, and therefore spacetime no longer a dynamic somewhat, what is the true nature of the void-vacuum? What is Ginnungagap? Does "consciousness" reside here?

No one can answer these questions, but the capacity to frame them illustrates a peculiar fact about consciousness and the possible role it plays in universal unfolding. It may be that that intimate yet indefinable somewhat — the self-centering illumination we know acutely as consciousness — may be the ultimate no-thing that grounds the whole universe. That what we know as bare awareness is the very ground of the universe out of which all things may come.

Without any substantial objects occupying them, space and time lose all meaning. Without *relation* or *interaction* between "somethings" situated within them, space and time cannot even be defined. This is why it may be by entangling quantum degrees of freedom or qubits, bare points in the quantum foam, that a universe may first come to be. Yes, it may be the formation of information that begets the primary step in the creation of creation. A possibility that also sheds light on dark matter and the holographic nature of Nature.[*]

When we see consciousness for what it is, as it is, we understand it as a kind of emptiness that possesses an inherent fundamentality. A formless no-thing that breathes life and meaning into all creation and as the container of opposites is itself beyond and without opposite. It is 'Sunyata' to use a Buddhist term. But we are getting ahead of ourselves, before we ruminate on the possible origins and invariant structures of conscious awareness, let's set the stage for a conceptualization of consciousness and novel ontology by returning to the scientific image and analyzing the QFT that illuminates the nature of light and explains how it interfaces with matter — quantum electrodynamics (QED).

But first, let us analyze some…

---

[*] See Carroll, S.: *Something Deeply Hidden.* (2019)

Ancient Insights

Over two thousand years ago, in disparate places around the earth, a common cognitive revolution took place. Rationality was rising out of myth, beginning to concern itself with proof, demanding reason—the heresy of questioning the Gods giving way to humanity's innate desire for truth and understanding. The Pythagorean theorem was the first inalienable argument that you could not refute, for, backed by mathematical precision, it offered *proof.* There could be no debate over the answer. Gifted to us by mathematical certainty, Plato and his famous interlocutors used this form of conceptual clarity to help fuel the evolution of Greece's use of rationality. The word itself displays a hint of its dependence on mathematical precision as (ratio)nality is a form of logic-based reasoning where things stand in relation to one another.

It can be seen in every formal sentence in language as it uses this subject-predicate structure. One could argue that it was Pythagoras' clarifying insight merging the nature of our reasoning with mathematical precision that allowed our culturally shared social reality to flourish. This brought new forms of government and justice systems, to trade and monetary policies and socioeconomics. All that from a simple insight.

Before Pythagoras of Samos, Thales of Miletus made some of the first salient observations about the nature of fields. Unbeknownst to him, he was the first to notice 'static electricity,' which is more deeply, the electrostatic *field*. He would charge a piece of amber — fossilized tree resin — by rubbing it on fur and it would thereafter attract or repel light objects. Like magic, it would lift a feather from a table. Aside from this small observation, he didn't know what to make of it. During his time, amber was known by the name "ēlektron," which shares an etymological root meaning 'shining sun.' Many years later, André-Marie Ampère would use this word to name his electrodynamic molecule.

Unknown to Thales, electrons are indeed the fundamental quantum responsible for the electrostatic effects he once observed and are responsible for the production of light, or 'shining the sun.'

Thales was also a material monist. That is, he believed everything in the universe was made from a single substance: water. The universe as a fluid may be a strange ontology but becomes fascinating if one views modern QFT as a theory of waves in fluids. A poignant philosopher to begin our descent into the quantum fluids of spacetime, so let's analyze his observation with modern understanding.

What he was doing, however, was activating and accessing an electric field that was already there. He was creating an imbalance of electric charges by supplying the amber with an excess of electric potential. The field it was generating reached out to affect other nearby electric fields. The electric field can be either an attractive or repulsive force. It just depends on the sign of the charge. That 'sign' a consequence of the direction of a quantum's phase shifts.

Consider, an electric charge sets up a disturbance in the underlying electric field. This disturbance moves out radially, and its effective "strength" grows dimmer with distance, just as with the light or gravitational force of the sun. Now, if one places another charge in the field created by the first, a force will be felt by both charges. Whether they repel or attract depends only on the sign of the charges: opposites attract while like charges repel.

The force felt is not the field but the "fluid" of the field. I share Faraday's view that the field is there and will always be there before any and all charge interactions. The electric charge merely "turns on" or accesses the underlying field. An electron is a local field disturbance, a radial dispersion that sets up the possibility for a corresponding force should another electron enter its vicinity and "feel" it.

This divergence from the charge can become a force if and only if another charge is nearby because the interaction between the two charges results in a "force," which is just the warping of electric fluid. The critical point to note about the field is that it's an omnipresent and timeless no-thing that only exists to allow; an attribute of the quantum foam.

Returning to the prehistory of light. Sometime later, still in Ancient Greece, a materialist-cosmogenic theory was put forward by the pre-Socratic philosopher Empedocles. He believed that the universe was made up of four basic elements: Earth, Air, Water, and Fire. He also adhered to an extramission theory of vision; that our eyes beamed out rays of light to see objects. For him, the goddess Aphrodite gave the gift of sight to man. By combining these elements appropriately, Aphrodite created the eye and lit within it a fire. That fire's light was thought to shine out from the eye to touch the objects outside of us so that we could see them.

The first to reject this view and argue for an intromission theory of vision was Aristotle. He would also remark on light and vision that whenever we get close enough to one another, we can see the reflection of ourselves in the eyes of others. As for the elementary constitution of the universe, Aristotle would introduce a 'fifth' element beyond the four of Empedocles: a "quintessence" that was light itself. Mythologically, it was known as the Aether, the substance of empty space, and air breathed by the Gods. For Aristotle, the Aether was ubiquitous, saturating the space in between the stars as darkness itself but of one of such a nature that it allowed for the transmission of light. Today he stands vindicated as we know this as the essence of the EM field itself.

As regards vision, later still, the immortal geometer Euclid, following the Empedoclean-turned-Platonic tradition of vision and its theory of rays emanating from the eye, would write a treatise called *Optics* wherein he discerned perspective.

It was once thought that our eyes beamed out to the objects we perceive, and that is how we saw. Only later did Johannes Kepler definitively show that the opposite was true. The eye is a *camera obscura* as light produces upon it a retinal image. Light is reflected from the object and makes its way to our eyes enabling us to see what is "out there."

This result, however, is not in line with our theory of mind. But one that recognizes both extramission and intromission theories of vision as regards consciousness, maybe the actual situation is more akin to the remark made by the Persian polymath Avicenna in the 11th century: "The eye is like a mirror, and the visible object is like the thing reflected in the mirror."

Consciousness does indeed "beam awareness" out to the objects of its concern, illuminating them for itself. When reading a book, for instance, one's *awareness* is not only within their head — unfolding language with a semantic decoder that is their prefrontal cortex — but it is also, equally, and actually, situated on the very page itself.

The En(light)enment

Around the eve of the Enlightenment, Rene Descartes — inventor of the Cartesian coordinate system and mind/matter substance dualism — stated that light was a property of the "luminous body" that came to us through a transmitting medium. Later, Newton — armed with this insight and a crystal prism — would discover that pure light from the sun carried within it every color of the rainbow. That despite his belief in its particulate nature, the only way it could refract into its various colors was because each color had a different wavelength. Light is a wave.

When it passes through a prism, every differing wavelength gets refracted slightly more or less than its neighbor, and the colors of the rainbow appear. Another fact of the matter illustrated by Newton's prism is that the different wavelengths of light occupy the same space. Not only can different wavelengths of light occupy the same space, but *identical* wavelengths can as well. When they do, they sum. It's this capacity of light (and of most bosons in general) that has allowed us to achieve light amplification by the stimulated emission of radiation. In a word, invent the laser.

The first physicist to rigorously argue for the wave nature of light was Christian Huygens but its proof was made by Thomas Young. For it was he that performed the first "double slit" experiment and proved without a doubt that *light is a wave*. To see what he accomplished, imagine a surface with two narrow slits and shine a light on it from a far enough distance such that the wavefront reaches the slits flatly. Upon exiting the other side, if the wavelengths of light are shorter than the slits are wide, the waves will diffract around the edges and propagate out spherically.

The now doubled spherical wavefront will begin to interfere with itself, and upon reaching the far wall, a pattern of bright and dark bands will appear. The bright bands are the result of wave crests interfering constructively and gaining amplitude, while the troughs of the wave interfere de-constructively and lose amplitude such that no light will arrive at the screen.

This simple experiment not only reveals the true nature of light as a wave but also illustrates a wave's capacity to superpose—that is, interact with itself. For even a single wave — a single photon! — will produce this pattern thereby illustrating its nonlocal spatial extension and capacity to interact with itself. To be fair, it is only after many "trials" with single photons that the pattern emerges. This is because photons interact at point-like locations, and one must gather the data from many trials before the interference pattern will materialize. Regardless, the pattern that always emerges owes its reality to a *wave's* capacity to *superpose*.

From Huygen's and Young's efforts, we understand that light is a wave in a medium. A medium once referred to as the "luminiferous ether." This ether, like the Aether of old, was thought to be ubiquitous. An omnipresent fluid, *at absolute rest*, filling the whole of the universe. A kind of stationary fluid that we, flying on our spaceship earth, must be *flowing through*. Thus, it was believed, if it exists and we are moving through it, we should be able to measure its effect and detect it.

The famous experiments of Michelson and Morley would later disprove *this* conception of a mechanical kind of ether. As it turns out, there is no such thing as a universally still water; an absolute frame of reference that is "at rest." It took special relativity to prove that spacetime itself has its own dynamics such that it hides the true nature of the Aether.

The "luminiferous ether" persists, it merely has different properties than once originally anticipated. It's not an unmoving substrate but a dynamic entity in its own right. Today, we call this new understanding of the Aether – the EM field. The luminiferous ether does not exist, but the EM field stands in its place with far more spectacular dynamics.

Still, we know from Descartes, Huygens, and Young that light is a wave. But a wave in what? How do we arrive at the understanding that light is a wave in a strange EM ether? The answer requires introducing more genii; one experimental and one mathematical. Michael Faraday and James Clerk Maxwell were able to turn this conception of a mechanical ether into a "vortex sponge" and arrive at the correct EM field equations and proper concept. How did they do it?

Curiously, having nothing to do with light *per se*, the catalytic insight was stumbled upon by the Danish physicist Hans Oersted. While lecturing on electricity, he noticed a compass needle move near a current-carrying wire. This happenstance seeded the inquisitive mind with the knowledge that there is a deep connection between electricity and magnetism.

Later, in 1845, working from Oersted's experimental insight, Michael Faraday was able to show its inverse — that a changing magnetic field could induce an electric current — further deepening our intuition regarding the interconnection between electricity and magnetism. The father of fields, Faraday was the first person to take seriously the notion of a property of "mere" space and in arguing for its physical, albeit immaterial, actuality, he developed the proper concept of a field. Faraday saw "lines of force" extending out from and returning to a bar magnet. These "lines of force" carry energy and momentum through "mere" space; the measure of which we now call magnetic flux. He later extended his field concept to electrostatics and the charges between them.

Before Faraday, no known mechanism existed that could propagate a change from place to place. The prevailing notion prior was called 'action-at-a-distance,' and no one understood it. Even Newton, armed with his monumental insights into the nature of gravity, knew not how its effect was generated and made manifest and famously remarked that instantaneous gravitational influence made little sense. In a letter to Richard Bentley, he wrote: "As to what the cause of gravity is, I do not pretend to know."

Nevertheless, it was only by replacing the fallacious notion of action-at-a-distance with that of a field that physics would precede. Today we are at a similar crossroads, only the concept in need of discarding is entrenched. So much so that many cannot even consider its abolition; it's the point particle, and it does not exist.

Einstein would later write of Faraday's insight: "A new concept appears in physics, the most important since Newton's time: the field. It needed great scientific imagination to realize that it's not the charges nor the particles but the field in the space between the charges and the particles that are essential for the description of physical phenomena.

The field concept proved successful when it led to the formulation of Maxwell's equations describing the structure of the electromagnetic field."˙

The discovery that light is a unified form of electric and magnetic phenomena will forever remain, until the end of time, one of the greatest intellectual achievements ever accomplished by our noble species. A timeless truth that existed antecedent to our advent and will remain unto eternity. The crowning achievement of all this intense research regarding electricity and magnetism was made in 1864 by the Scot, James Clark Maxwell. By his genius he made a timeless discovery and was rewarded for his effort with an insight whose phenomenological experience must have felt like an indescribable bliss.

By discovering what light is, Maxwell encountered an angel. Not a literal angel, of course, if there be such a thing, but a metaphorical one. To uncover her, he had to internalize countless previous insights, involve constants and measurements made by Faraday, Ampere, Coulomb, and many others, learn a new kind of mathematics (partial differential equations as they involve rates of change concerning continuous variables), and supply a "law" of his own. Again, differential equations are needed to specify the intensity of the field at every point in space and to describe how that intensity changes over time and see the effects it has on adjacent areas of the field.

The result was a set of four of the most potent equations in physics, and a secret lay within them. A constant emerged from them, which he denoted by the letter c, possibly because *Celeritas* is the Latin word for speed. His equations used experimentally verified quantities known as the permittivity and permeability of space. Upon calculating the rate at which changes in the field would propagate, a timeless insight was revealed to him. He discovered the speed by which electric and magnetic fields co-produce one another and spread through the EM field was the speed of light.

In his own words, Maxwell writes: "The theory I propose may... be called a theory of the Electro-magnetic Field because it has to do with the space in the neighbourhood of the electric or magnetic bodies, and it may be called a Dynamical theory because it assumes that in the space there is matter in motion by which the observed electro-magnetic phenomena are produced." And "What, then, is light according to the electromagnetic theory? It consists of alternate and opposite rapidly recurring transverse magnetic disturbances, accompanied with electric displacements, the direction of the electric displacement being at the right angles to the magnetic disturbance, and both at right angles to the direction of the ray."[†]

---

[*] Einstein, A., Infeld, L.: *The Evolution of Physics*. Cambridge University Press. (1938)

[†] Maxwell, J, C.: *A Dynamical Theory of the Electromagnetic Field*. (1864)

Today, we know that the EM field saturates all of space and time. It is the Aether of Aristotle; the air breathed by his mythological gods. One need only imagine placing their eye way out in intergalactic space, and you know that you'd see a beautiful cosmic web of galaxies. Starlight from every conceivable direction, radiation in the EM field is there, yet this waving EM ocean is not yet light. For it to become light as we know it, it must acquire another feature: it must transmute itself into the luminescent and knowing quality that is the light of consciousness itself. For it there is no witness, the waves remain there, passing and flowing through one another without any friction, but they remain only lifeless fluctuations in a field.

Found in Maxwell's journal, we encounter a compelling passage that encapsulates the timeless beauty and initial ineffability that renders an eternal insight; he writes: "Happy is the man who can recognize in the work of to-day a connected portion of the work of life and an embodiment of the work of Eternity. The foundations of his confidence are unchangeable, for he has been made a partaker of Infinity. He strenuously works out his daily enterprises because the present is given him for a possession.

Thus, ought man to be an impersonation of the divine process of nature, and to show forth the union of the infinite with the finite, not slighting his temporal existence, remembering that in it only is individual action possible, nor yet shutting out from his view that which is eternal, knowing that Time is a mystery which man cannot endure to contemplate until eternal Truth enlighten it."[*]

We have come to know that light is a co-producing oscillation of electric and magnetic fields that travels at a specific speed. This speed is due to the elasticity of the field it oscillates in as all fields possess a kind of cohesion that allows the waves within them to propagate; for one spatial point to affect its neighbors. The *ease* with which a field allows this is called its elasticity, and if the field is a unified "thing," it must be same everywhere and therefore, *specific*. It can only be one way and in the case of the EM, we know it to be c.

If a field lacked elasticity, its waves — light (or photons) in the case of the EM field — would travel at infinite speed, and nothing in the universe would make sense. A consequence: the elasticity of the EM field sets a universal speed limit and consequent "rate" of causality. This restricts the swiftness by which information-carrying signals can move through the universe and can now be seen as a *requirement* for coherency.

Quantum nonlocality — a process we will later consider in detail — does not violate this principle as it cannot be used to communicate information

---

[*] Maxwell, J, C.: Letter to Frederic Farrar. (1854)

superluminally. Regardless, in the EM field, every photon is a massless wave that will propagate at this speed as it cannot ignore the elasticity of the field. It matters not what set the wave in motion, how it began vibrating or what its amplitude and wavelengths are; all EM waves move at *c*. As we shall see in due time, matter-waves like electrons, unlike photons, possess a characteristic that allows them to couple to other fields, thereby creating different potentials. This slows them down and confers unto them inertia or mass.

Aristotle's Aether became Maxwell's angel, and today we understand the darkness of space that carries light, indeed is light, is a real thing: the EM field. Although it's not the air breathed by gods, nor a 'vortex sponge,' nor is it a frame of absolute rest possessing the mechanical properties it was once believed to have, the EM field is nevertheless a 'thing.'

The Aether — now understood as a real physical "object;" a field — became a steppingstone of complexity laid bare for following genii. The next great revolution would come at the turn of the century when a lone genius would initiate three pillars of physical theory.

In the 'miraculous' year of 1905, a young Albert Einstein would validate statistical methods in physics and prove the existence of atoms, solidify quantum theory by using Planck's constant *h* in his work on the photoelectric effect, and finally, would decimate our Newtonian conceptions of space and time with his theory of special relativity. The cherry on top is the most famous equation in all of physics; by employing the newly discovered structure of space and time, Einstein would go on to prove mass-energy equivalence. However, Einstein would not go on to explain the origin of the field. That discovery would take much longer and more work in beautiful yet seemingly useless mathematics.

They are known as Yang-Mills or Gauge theories, and due to them, we understand the EM field as a necessary *interaction* field. The name "gauge" theory is a historical relic. The proper name for these fields, were they to be named with today's understanding, would be a *phase-field* as these fields owe their origin to the *phase* shifts of certain elementary quanta. For instance, in order for electrons to interact with one another, they must do so through an intermediate field. They do this by "passing" virtual photons back and forth. As such, the field itself is required for its matter quanta to interact and is an essential consequence of nature's need to keep its geometrical and symmetrical properties locally coherent.

This mechanism, the method by which nature produces a force and keeps track of its evolution is called **non-abelian phase symmetry**. It refers to the order of operations when considering the rotations and phase-shifts of quantum waves and how these movements communicate with other parts of space. In other words, to produce virtual EM field vibrations that appear as a force, electrons must rotate and shift phase. These transformations of the electron create photons that propagate the change away, taking the information about the shifts of the electrons to other parts of local space. Thus, the movements of matter quanta — fermions — *generate* waves in a *necessary* connection field between them. All QFTs possess this structure. For the electroweak interaction, the necessary phase field is the EM field. For the strong interaction concerning quarks, the basic idea is the same, but the connection field(s) are color gluon field(s), but these have much more complicated dynamics as the "symmetry group" is much larger. In the strong interaction — unlike photons that don't respond to one another — color gluons do feel the presence of other gluons *and* quarks. But again, we are getting ahead of ourselves.

Let us return to Einstein and Maxwell and see how his placing restrictions on his angel led to a completely novel conception of space and time.

"...the simplest of all natural phenomena, namely gravity, which does not cease to strive and press towards an extensionless central point, whose attainment would be the annihilation of itself and of matter; it would not cease, even if the whole universe were already rolled into a ball." - Arthur Schopenhauer

CHAPTER IX

Invariant Angel | Malleable Spacetime

Proven by his equations, the timeless insight awaiting Maxwell regarding the speed at which his angel takes flight was next used by Einstein to form the special theory of relativity. It is "special" because it restricts itself to constant, rectilinear motion. In fact, the entire edifice of special relativity rests on just two postulates, and from them, mind-blowing consequences emerge—strange phenomena like length contraction, time dilation, and the non-simultaneity of events.

The first postulate — a "law of nature" — dropped out of Maxwell's equations; the unalterable, unchanging speed of light. It's a "law of nature" because it's the same for everybody. Anyone can perform the experiments and calculations in their laboratory (reference frame) and arrive at and measure the constant c. Light travels at a specific rate in the vacuum, a consequence of its elastic properties, and so dictates how a wave in it *must* propagate. As we know, *any* disturbance set up in this field will move at c, there is just no getting around it.

The second postulate, known to Galileo but promoted to a meta-law of nature by Einstein — that is, a law about the laws of nature — was **the principle of relativity**: all laws of nature are the same for everyone no matter which frame of reference you find yourself in. If one sets up a laboratory inside a black hole, on the surface of the sun, or in a rocket-ship traveling near the speed of light, nature's laws will be the same for every one of them.

When Einstein began to piece together these two postulates — the invariant speed of light and that it matters from whose frame one is observing — Newton's Cathedral of Absolute Space and Time crumbled. If the speed of light is to be the same for everyone — every freely moving observer — naïve notions of distance, time, and simultaneity must all be reconsidered and understood as *relative* to each observer and their particular frame of reference. Again, it matters who is looking, and the answers one gets will be relative to the frame of reference one is in.

Phenomenologically, we know a frame of reference rather intimately as "the self;" the zero-point seat of conscious experience. Physically — as not all physical systems need to be aware — a frame of reference is simply a specific coordinate in a spatio-temporal grid. Frames of reference split from one another and mix when one adds a novel ingredient; motion. Einstein discovered that motion — communicated information in the form of light — *mixes* space and time. To keep the speed of light constant, space and time must conspire together, ebb and flow, and borrow the other's intrinsic aspect to do so.

It's a fact that there exists no preferred direction in space, and "motion" is something that can only be discovered by comparing one's state to that of something else; it's purely relative. It's the *difference between* frames of reference where all the curious phenomena like length contraction and time dilation take place. This "place" is defined by something called a **Lorentz transformation**. A Lorentz transformation describes how the coordinates of separate reference frames transform into one another. Dilating time or contracting space so the speed of light can remain absolute.

The symmetry underlying a Lorentz transformation involves three parameters: translations, rotations, and boosts. Once all three of these conditions are understood and adequately interrelated, one can see the difference between frames. One might ask the question: "Well then, what exactly does light do when we move from one frame to another?" The answer is simple: nothing. The speed of light remains constant between frames; it's the spatial and temporal units that must bend to accommodate light's unalterable speed. Hence, length in one frame may contract and have its time dilate so that the other frame doesn't notice a difference. In fact, no matter which frame one is in it will "remain the same" as seen by them.

A simple way to acquire an intuitive glimpse into special relativity is to compare the time measurements of two different observers. But first, how does one measure time? Despite many ancient cultures' efforts to do so, by carving rock calendars and building sundials, it wasn't until the 16th century that Galileo came up with an almost perfect way to measure time. Galileo showed how a pendulum's sway oscillates at a near-constant frequency, and one could count periodic increments or 'measure time' by its hypnotic motion. This "harmonic oscillator" seeded the pocket watch and clocks in general. But still, these clocks are too bulky and rudimentary for our purposes, so let's fashion an *ideal* clock.

Imagine two mirrors a certain distance apart in empty space where a photon bounces between them. One trip to and fro is called a period. With the distance between the mirrors and the velocity of light unchanging, the clock is perfect. Now, we duplicate our clock and give one to Alice and another to the Hatter. Sitting side by side, their clocks are synchronized. What happens if one is given a boost and begins moving away at a constant speed? It does not matter which because with no other frames to deduce who is moving, each will see the other as moving and themselves as stationary. From the perspective of Alice's reference frame, she will see the Hatter's photon travel a greater distance to complete a period as it now travels diagonally.

However, the Hatter himself does not see this effect; by staying with his clock, he observes the photon oscillating straight up and down; for him, nothing changes. But we know that the speed of light is constant, and is in one frame traveling farther, so it must take more time. But we know that's not right because we share the same reality. So, what must be taking place to keep reality coherent? The answer: the Lorentz transformation. "Between" the frames, space and time are borrowing one another's attributes to maintain light's constancy.

To keep light's locomotion constant, each observer will observe the other's clock as moving slower. In the jargon, their time will dilate.

The main takeaway from special relativity is that constant, relative motion mixes space and time. What occurs is that space borrows from time, and time borrows from space, and this exchange keeps reality consistent. Superseding Newton, neither time nor space is absolute. Instead, they are dynamic, fluid-like entities.

Pearls of Aether-Consuming Onyx

In the years that followed, Einstein wished to "generalize" relativity to include gravity as well as curved and accelerating motions. He knew, along with Newton, that gravity's supposed "instantaneous action-at-a-distance" was not quite correct. But above and beyond Newton, Einstein knew that this "spooky action-at-a-distance" violates the principle of relativity. In Newton's theory if the sun were to disappear the earth would also at that very moment cease to be caught in its orbit. In dealing with relativity and its effects, one must ask which frame of reference we are considering.

To the elation of Einstein, Max Planck too would become enamored with the ideas of special relativity. In 1910, Planck would say of his special theory of relativity, "if it should prove to be correct, as I expect it will, he will be considered the Copernicus of the 20th century." Planck could not have been more correct, and Einstein wasn't finished. A year later, Einstein would put forth his **principle of equivalence**. This equates the uniform motion involved with Galilean transformations with the non-uniform motion of acceleration.

Basically, it shows that gravity and acceleration are indistinguishable from one another and thus are equivalent.

To arrive at this idea, Einstein had what he later called the happiest thought of his life. In his words: "The gravitational field has only a relative existence... Because for an observer freely falling from the roof of a house - at least in his immediate surroundings - there exists no gravitational field." The true beauty of this insight is that it was arrived at phenomenologically. As an embodied being, Einstein's eureka moment barreled in on the wings of a "what it feels like." Or in this case, a "what-it-*would*-feel-like."

Now, what could he mean by exclaiming gravity has only a "relative" existence? Consider the "vomit comet," to gift its passengers with the *feeling* of weightlessness, for a few thousand dollars, one can board a seatless airplane that completes massive parabolic arcs high in the stratosphere. In this plane, we recognize that when we can respond freely to gravity its field of force vanishes. Poof! When the plane must regain altitude, it returns the feeling of weight to its passengers, adding a little extra.

This is why physicists do not discern the difference between centrifugal (acceleration) and gravitational force. Again, because they are indistinguishable from one another. Not to belay the point but to reveal this more clearly, consider the training army pilots must endure to fly their billion-dollar, death-delivery machines; G-Force training. G-force, of course, stands for gravitational force but for these pilots, it is brought about by centrifugal means. A massive centrifuge spins the pilots around at face-deforming speeds, creating "gravity" out of nothing but acceleration.

It's a testament to Einstein's genius that he found the clue to general relativity in these artificial fields of force and see that the sensation of weight is not felt when we are free to respond to gravity. Gravity only "appears" when there exists something to prevent our falling. This deep insight told Einstein that gravity was something deeper than merely a force. It must be embedded within, intrinsic to the very fabric and structure of reality.

From here, he would need only to decipher the complex mathematics of Riemann's non-Euclidean geometry and apply it to the cloth that is Minkowski spacetime. To do this, he had to add a property to spacetime itself. For the objects in spacetime to remain symmetrical after undergoing transformations, space and time themselves have to become fluid. Not just a Lorentz transformation kind of fluid, but a warping geometry superfluid.

Consider for the moment anamorphic art, a perspective technique that seems to show a distorted image. However, when the work is viewed from the appropriate angle the distorted object is seen as undistorted. Similarly, consider your image as it appears in a rounded doorknob as an honest and accurate representation of "how things are." In generalized, curved spacetime, both you and your image in the doorknob are equally valid frames of reference that capture one and the same thing. This is what the space-filling fluid of general relativity "does" to spacetime.

To arrive at this understanding of "curved" spacetime was not easy. For three years, Einstein was tutored in and struggled with the mathematics and in the fall of 1915, he was ready to publicize his work. The first test of which was to calculate the perihelion shift of the tortured marble Mercury. When the wobble of Mercury agreed with Einstein's gravitational field equations, he was overjoyed and unable to work for days after, overcome by the epiphanic bliss that accompanies eureka insight.

His hard work showed that gravity is the curvature of the underlying structure of spacetime itself. It's not a *force* per se but is instead a geometrical consequence of the presence of condensed energy-momentum concentrated and swimming in spacetime. Matter is responsible for the bending and warping of spacetime and we feel this malleable transformation as gravity. The now common adage goes like this, matter tells spacetime how to move, spacetime tells matter how *it can* move.

Like all great scientific achievements, general relativity contained more than an oceanic spacetime. Further analyses of its formal structure continued to reveal novel concepts as well as a seemingly impossible object. A "geodesic" is the "straightest" line path an object can travel through curved spacetime. A photon, although massless, has non-zero energy and is thus affected by gravity. As such, it follows a geodesic as it travels the straightest line possible through curved spacetime. This idea gives to the photon a false image as a kind of beam or ray, but we must remember, a photon is a wave, and we are only able to "see" the path of its travel after it has collapsed into the apparatus that registers it.

Nevertheless, this "bending of light," which is really seeing its necessary relation to spacetime itself, was one of the first experimental confirmations of general relativity. In 1919, the astronomer Arthur Eddington traveled to South Africa to observe a solar eclipse and see if light bent around the sun. In measuring the positions of distant stars behind our sun, he was able to confirm that the gravity — the total energy-momentum — of the sun bent the surrounding fabric of spacetime, guiding light in how it should move. Today, cosmologists call this effect gravitational lensing. An effect that would be inexplicable if it were not for general relativity.

As for its "impossible" object, GR has predicted the existence of pearls of Aether-consuming onyx, known more commonly as black holes. These are orbs of impenetrable darkness that crush space and time and insatiably swallow light. They are awesome.

General relativity can be understood by the light of two equal images: as curved spacetime with fluid-like transformations, or simply as a *metric field*. It represents the mathematically flat but malleable-by-matter fabric of space-time. The mother of all fields dictates the possible behaviors of all the others superimposed within it. In it, the idea of a metric fluid replaces the idea of gravitational force, for there is no force, just the ebb, and flow of spacetime's energy-momentum transformations.

Finally, considering the cosmos as a kind of ocean allowed us to prove the existence of gravitational waves. Fluid compressions of the fabric of spacetime have now been experimentally observed. In 2014, the team at the Laser Interferometer Gravitational-Wave Observatory (LIGO) performed a kind of upgraded Michaelson-Morley experiment and observed the contraction of space caused by two black holes colliding more than 1.3 billion light years away. Two L-shaped buildings, four kilometers long, but three-thousand miles apart, each measured a contraction of space to an accuracy of 1/10,000th the width of a proton! It was able to show that one building was contracting along its length while the other remained the same. This also confirmed the direction of these transverse waves. As one of humanity's most accurate experiments, LIGO is one of our greatest achievements. It stands as a testament to the power of the insatiably inquisitive, collective, rational mind.

By now, we have well-established the baseless fabric of this vision. We have seen that light is a wave in a field and that GR contains waves that flow through the metric field which is spacetime itself. But what of matter? How is it a wave? And what of individual, separable systems? We are now able to consider the quantum.

"...in the mystical feeling, the truth is apprehended from within and is, as it should be, a part of ourselves."* - Arthur Eddington

CHAPTER X

The Quintessence of Quanta

Quantum mechanics is commonly thought to be a theory that describes the inner workings of the subatomic microcosm, the realm out of which the very furniture of reality begins to take shape. It is thought to be a theory of tiny, too-small-too-see particles that gather to form matter as we know it. And as a theory of subatomic particles, it's known for its audacity and many paradoxical phenomena, e.g., such as superposition: the phenomenon of a single object existing in two (or more) places simultaneously.

On the face of it, it can seem so counterintuitive that many make the mistake of believing that the quantum realm is somehow separate from the "classical" world of our everyday and can thereby play by a different set of rules. Although the stratification of reality into sedimentary "layers" is a helpful conceptual tool, e.g., atomic, chemical, biological... the idea that there is a "quantum realm" existing 'beneath' the world we commonly encounter is simply misguided.

But the intuition behind this line of thought is not for want as it truly does locate a "place," a stratum that marks the quantum-to-classical transition. This "place" marks where quantum effects dissolve and the classical world of our everyday experience emerges. How the quantum achieves this will be analyzed in what's to come. Know now that the first step involves discarding the idea that quantum theory deals in and with particles; it simply does not.

Reality is quantum through and through, and what it tells us is that hiding behind the façade of reality's fortified forms and firm furniture is instead an oceanic, immaterial world of energetic, resonating waves that endlessly ebb and flow through and intermingle with one another. Again, the common conception that quantum theory deals in sub-atomic particles is simply wrong and calling it a mechanics is nothing but a misnomer. The theory does not define the cause-and-effect relations of solid bodies in space — particles as the

---

* Eddington, Arthur.: *The Nature of the Physical World.* (1928)

well-defined gears and cogs that populate classical mechanics — and the forces placed on them.

Instead, QFT provides us with a different notion of base reality altogether, describing a kaleidoscopic tapestry woven of immaterial fields brought to life by their resident quanta of energy. QFT is not a theory of the subatomic microcosm but is instead and is more accurately a theory that describes the very fabric of the cosmos.

To understand the theory will require a deep understanding of its constituent objects: mainly, quanta and fields. Quanta are waves of energy that flow in fields. While fields are omnipresent and unbounded determinable properties of space. Fields may be more easily intuited metaphorically, as ethereal waters that saturate the cosmos, distinct fluids that each that each possess inherent characteristics. When a field entertains a unique disturbance of itself, that is, a quantum of energy, the ebb and flow of this motion is expressed as a wave that actuates and activates said field's property. All fields share a common and unified carrier sometimes referred to as the quantum foam.

In this picture, particles no longer stand out against an empty vacuum of nothingness, but rather space itself is revealed to be an oceanic plenum—a timeless, sentient sea where everything is always-already interconnected. In fact, at a certain level of focus, the quanta that populate fields are indistinguishable from one another. Thus, the world is an indivisible whole, and we, its cognitive agents, are unknowingly ensnared and equally entangled in an invisible, multi-dimensional cosmic web.

With this understanding, quantum mechanics integrates seamlessly with the larger framework of reality, painting a coherent and unified picture of the cosmos.

☾

Borne out of the endless effort to understand the deepest workings of nature; quantum field theory has become our most precise, experimentally verified theory. As with any physical theory, however, it has been plagued by problems and suffers from many interpretational anomalies. Not only is QFT's theoretical predictions some of the most accurate and agree with experiment to an unrivaled degree, but its very dynamics predict some astoundingly strange and counterintuitive phenomena. If taken verbatim, its core equation predicts an endless bifurcation and proliferation of equally real but unable to interact worlds. A new universe created every time a "measurement" (collapse, interaction, decoherence) occurs.

The resulting ontology implied — Many-Worlds — would put the Norse world-tree Yggdrasil to shame. Although we can allow ourselves to be led by

The Quintessence of Quanta

the mathematical formalism, we do not have to adhere to its every implication. Nevertheless, there exist seemingly irresolvable problems facing the theory. Experimentally, we must contend with quantum "jumps," "wavefunction collapse," and "tunneling," to name but a few. Theoretically, the mathematics requires **renormalization** as the solutions to some of its calculations arrive at infinity. Some are not happy with this scenario. And again, there is always the underlying and contradicting ontology of wave versus particles. Given all these unsatisfactory implications, how are we to come to grips with what the theory is trying to tell us?

Luckily, after a hundred years since its inception, an image is beginning to come into focus, and it forms a radical departure from the classical mechanics of old.

Unfortunately for Democritus, one of the most surprising results to arise from QFT is that rock-bottom, bits of matter do not exist. However, he was not wholly incorrect. His "philosophical" atom — a "smallest" unit of substance — *does* exist. But it is a fragile energy-wave, not a solid "piece" of matter.

In the heart of matter, we find no matter. Quantum theory has gifted us a new understanding of what a subatomic unit is, and it is most certainly not a dimensionless point-particle floating in empty space. Empty space doesn't exist, and when quantum mechanics and special relativity constrain what a particle can be, they are doomed to dust. Particles simply cannot adhere to the requirements placed on them by both theories. If one holds that particles are the essential reality, they will forever live in confusion.

If one holds a particulate view of quanta, wave/particle duality, quantum tunneling, superposition, nonlocality, and entanglement all make little sense. Ontologically speaking, no "thing" can be both a wave and a particle at the same time. It is one or the other and particles are not up to the task. A consequence of logic alone, no solid particle can tunnel through an equally solid barrier, be in two places at once, or instantly affect and superluminally communicate with another spatially separate other.

But if one understands that an always-already nonlocal (spatially extended) *wave* — a dynamic shape of the field it is in — form the basic reality, then all of these counterintuitive notions are brough into light, and the paradoxes generated by them are lost. For instance, to "tunnel," a quantum as wave of field energy begins to encounter a barrier — which is itself made of stationary, geometrically-bound, entangled waves and is better understood as an energy-momentum *density* of a field rather than a solid wall — and some of it is reflected while other "parts" of the wave begin to interpenetrate the "wall." As this nebulous "blob" of field energy encounters the wall, probabilities of certain outcomes begin to climb. The wave can reflect, collapse, and interact to deposit its energy, or possibly "slide through" a gap in electric field density between the barrier's molecular bonds and emerge from the other side "as if" nothing had happened. Given the wave nature of quanta this is the only plausible explanation. The shape of the waveform and its attributes will have been affected by the transition and an examination of the wave after it has tunneled will reflect that.

It is even most likely that *some* of the wave was reflected back while another portion of it made it through. Until a measurement is made on it to find out where it is, the single unified wave is in a superposition of having gone through and not gone through the barrier. It is both at the same time, an easy "thing" to picture when one holds to the wave nature of quanta. Finally, because of its unity as a single "thing," — engendered by being a specific amount of energy — it must behave as one, and therefore its wavelike extension will collapse upon measurement.

If we follow the experimental results, led by but not beholden to the math, while allowing for some strange but observed behavior, and recognize that the wave's phase plays one of the most significant roles in QFT, then we are left with seeing the true object of these fields, waves. Albeit extremely complex waves that can interact with themselves (superposition), exhibit nonlocal effects, rotate, spin, and shift phase and because of how they interact with other waves behave *as if* they were particles. To be fair, the quantum as object supersedes its wave *and* particle precursors but the fact remains that quanta are best intuited as complex waves that possess two particle-like attributes. These are, they localize into semi-well-defined places when they interact, and they are countable — that despite their non-local spatial extension, they are unified across themselves.

At this point, one may wonder how superposed, spatially extended, ephemeral waves of energy become atoms? The answer, as we shall see, will cause us to deeply consider just what an atom is and of how waves behave when they are bound by some geometrical constraint.

As a preliminary precursor, QFT's conception of the atom sees nuclei — protons and neutrons — as a dynamic dance of energy flowing in quark and gluon fields, which, once stably formed, capture and cloak themselves with wandering electrons. The once freely floating electron becoming a standing wave pulsating about the atom, camouflaging the nucleus in a cloudy haze of electrical energy. Now a bound, standing wave, the electron exerts a well of electromagnetic influence into its local environment. This affects neighboring atoms, inviting them to form molecular structure or repelling them.

We should like to consider this image of atomic nuclei cloaked under the fog of an electron cloud as a 'somewhat' solid object, but this is not yet the case. This atom is as a composite object — involving a plethora of quanta and their resident fields combined — and will *still* behave like a wave in a field. This atom, so conceived, is not yet a classical "solid body in space" but is instead a transiently stable well of influence, a "shape" of its oceanic host medium.

Double-slit interference experiments bear this out. At the level where atoms are atoms and molecules are molecules, they retain and exhibit the nature of that which makes them up, quantum *waves*. The double-slit experiment has been done with mesoscopic objects; objects much larger than simple atoms and molecules and has convincingly shown that they to are waves. The question is then, if Nature is quantum through and through, how does it produce the macroworld that we inhabit?

The world of everyday experience is not wavy, it is classical because it is populated be "solid bodies in space." We must somehow make the quantum-to-classical transition while maintaining a coherent quantum theory. How are we to achieve this?

The answer: a mechanism called decoherence. Decoherence is a collapse-like quantum process that hides an unspoken philosophical commitment to realism as it illustrates that the environment itself serves as its own monitor. No conscious observers are necessary as the cosmos is constantly grounding itself by tracking and registering the interplay of its relata such that we can affirm the fact, barred from us philosophically, that physical world does indeed exist apart from us. Reality is real.

Decoherence serves the function of localising open quantum systems by "closing" them. It achieves this by collapsing quantum superpositions such that as material objects grow larger, they are made to entangle more and more with their enveloping environments. The result is the formation of a quantum system that is essentially no different than a classically closed system, namely, a "solid body in space." Furthermore, decoherence has it that information about whatever object — whether a molecule, rock, or organism — is irreversibly dispersed throughout its local environment.

If this all sounds a bit much, it's because I've been skipping ahead. Later, we will examine decoherence in detail but for now, let's go back to the start, dispense with the concept of dimensionless point-particles, and upgrade our conceptual understanding to that of the quantum.

The Quantum

A quantum is a highly unified, specific quantity of spatio-temporally extended field energy. It moves and evolves as a nebulous and dynamic wave, a shape of a fluid-like albeit "airy," immaterial field. Quanta are not "waves of probability" or particles, nor will they ever become particles. They are more akin to ethereal bubbles that express a field's characteristic. As true waves, they evolve into nonlocal superpositions.

Although they are not particles, they do possess two particle-like aspects. They are discrete, countable objects that may localize or superpose when made to interact. As waves begotten by a specific amount of energy, quanta are nonlocally unified, making them and their action "all or nothing" entities. It is this unity and wholeness that grant to them their unique abilities and leads to their appearance as particles. However, it's essential to understand that appearing-to-be and actually-being are mutually exclusive.

In the mathematical formalism, the description that captures all the information concerning a quantum system – its state, properties, degrees of freedom – is called a wavefunction, symbolized by the Greek letter psi (Ψ). The wavefunction maps space and assigns a complex value to each spatial coordinate that it occupies.

The evolution of quanta into superposition states raises an ontological problem about the wavefunction's actual representation. Called the **ontological problem of the wavefunction**, it asks what kind of causal map does it provide to correspond to an actual field configuration? How does it relate to the ontological modalities of actuality and possibility?

When the wavefunction as a mathematical object overlaps with, maps, and latches onto the physical reality of a quantum system, it is said to be Ψ-ontic. In this respect, the terms 'quantum' and 'wavefunction' denote one and the same thing. If the wavefunction is taken to represent only a mathematical abstraction and not how the quantum is in and of itself, it is said to be Ψ-epistemic. A Ψ-ontic quantum system is one which encompasses its possibilities as actualities.

As waves that misbehave, quanta exhibit peculiar behaviors such as "jumping," "tunneling," "teleporting," superposing, entangling, decohering, collapsing, and "spinning"—all of which lack classical analogs. For instance, in superposition, a unified quantum can exist in multiple configurations or states simultaneously, moving in different directions. As waves, quanta do not follow well-defined trajectories but explore all possible paths until they inevitably collapse upon reaching their destination.

Perhaps the most bewildering trait of quanta is their ability to collapse. Due simply to their unity as objects, the notion of collapse captures the fact that quanta can reconfigure their states instantaneously. This capacity is seen by their ability to (randomly) perform a certain amount of work (defined as the energy that they are) at any location of their nonlocally extended being. A process which simultaneously causes the quantum to re-localize itself.

When it does work, say, by interacting with another quantum system, its state or wavefunction is said to "collapse" such that it assumes a more localized configuration and it is this mechanism that reveals itself as one of the quantum's particle-like aspects. No quantum is safe from collapse because this process represents the localized points where nonlocal waves make themselves known to each other by interacting. Indeed, collapse and interact are, in some ways, synonymous terms. Also, it is worth mentioning, collapse has less to do with the inherent properties of quanta but instead has more to do with the things they come to interact with. A notion will shall consider in detail when we analyze the double-slit experiment.

The astute reader may have considered the notion that the instantaneous action of collapse seems to contradict SR as it picks out a preferred frame of reference and simultaneity slice of time. As such, novel dynamics must be introduced into the theory to relativize the simultaneity structure of Minkowski spacetime. As this is indeed possible, I will continue using the language of "instantaneity" when speaking of digitized — that is, discrete — quantum processes.

Concerning collapse, what is truly mysterious is that when it takes place, the quantum instantly vanishes from everywhere else it had been before the event. This is represented in the mathematics by literally discarding a portion of the equation. For many, this is hard to accept. Even though this inexplicable "spooky dynamic" is seen in experiments and is *implied* by many parts of the theory, its appearance forms the basis for its many philosophical "interpretations."

Put simply, at the schism between the theory's predictions and experimental outcomes lies a simple disagreement. Mathematically, the core equation and engine of quantum evolution — the Schrödinger equation — endlessly evolves into superpositions, never to arrive at the well-defined definite outcomes we retrieve from experiments. That we only ever observe a single definite outcome *requires* that we add the collapse postulate into the dynamics of Schrödinger evolution.

Again, one might think that the observation of well-defined outcomes in experiments entails that quanta are particles in the sense of point-like objects having had well-defined properties. Although this is not the case it does help to point out that quanta are only ever coherent units of field energy whose wavelike nature ensures they embody constantly vacillating properties. And as we shall see, is one which also aids to intuitively explain Heisenberg's indeterminacy relations. Furthermore, unlike particles that can retain their identity when they instantiate the same system, when quanta superpose and entangle to form a multiquantum system, they lose their distinguishing characteristics such that one can no longer tell one from the other. Thus, another defining feature of quanta is their indistinguishability. Whereas particles of old can be individuated within a system and retain their identities, quanta, on the other hand, can only aggregate into a unified system whereupon they lose their distinguishing traits by contributing to the novel whole. It is this mereological capacity that will allow us to see organisms in a whole new light.

The Quintessence of Quanta

To see their indistinguishable nature intuitively, consider two people holding opposite ends of a rope and have each shake it up and down such that a wave travels toward the center. Here, the two lumps will interact with one another, superposing and doubling the amplitude only to carry on and leave the interaction as if nothing had taking place. It looks as though they have traveled right through each other combining their energies in the middle. But is there another way to interpret this? Yes, it could equally be the case that where they superposed, they actually bounced off of one another. After an interaction like this, there's no way to tell which is which. On another note, an interaction like this will entangle the two lumps such that interacting with one will affect the other. But as is particular of my writing, we are once again getting ahead of ourselves.

At least now we know, a quantum is a global whole, a unified system and increment of energy that evolves nonlocally in the form of a complex wave, as a dynamic "shape" of a field or fields.

To better internalize these insights, let us now take a semi-historical detour to consider how we got here.

A Brief History

The word 'quantum' meaning 'how much' or 'how great' was first used by its discoverer, Max Planck, who, in 1900, solved "the ultraviolet catastrophe" by introducing it into his math. Planck's concern at the time was with the energy exchanges between a 'blackbody' and its production of thermal radiation — a transformation that turns matter into light (okay heat energy but you get the point). Theory at the time predicted infinite energy when up in the ultraviolet part of the spectrum. Which, if one is solving mathematical problems related to physical theory, an answer arriving at infinity is tantamount to a catastrophe — hence, the ultraviolet catastrophe. With common intuition about waves and the vibrations that produce them, Plank's initial desire was to explain the blackbody's radiation production via a continuous mechanism. But all attempts to do so failed.

In what Planck himself called 'an act of sheer desperation,' he introduced the quantum hypothesis and showed that a discrete, minimal amount of energy was being emitted, a "piece" or "unit" that could not be divided further. His' act of sheer desperation was to switch from a continuous range of values for the quantum's energy to a discrete one.

What is quantized is *energy*; it can never be less than a particular amount, is integer-valued, and can only rise in discrete steps. The formula Planck arrived at ($E = hf$) allows one to find the energy of a quantum by multiplying

The Quintessence of Quanta

its frequency (in Hertz) by his new proportionality constant *h* (measured in Joule/seconds). Thus, a simple relation arose between the energy of a wave and its frequency, a relation governed by what is now known as Planck's constant immortally symbolized by an italicized *h*. The magnitude of this new term is a specific amount, 6.6×10^{34}, and it silently implies a new kind of dynamic: instantaneous action. In physics, an action is a value that describes how a physical system changes over time and in Planck's case the action has *no* value; it is *timeless* so to speak. The discrete specificity of *h* also tells us is that a quantum wave is a unified thing, an object that will maintain its coherency and unity no matter how spread out in space it evolves to become.

Although its frequency remains relatively fixed, the possible locations that a photon, as "lump" of energy, can interact, spreads out as its wavefunction does. Where it will deposit itself as a unit of work performing function, it knows not. The spatial extent of the photon's wavefunction could evolve over billions of years, taking all-paths and becoming unfathomably immense. This is another reason why quanta aren't even necessarily small as their wavefunctions can grow, given enough time, to cosmological sizes. Still, when caused to interact, it will always deposit itself into a single awaiting atom—an atom that could be in your eye.

Let me restate that; no matter how spread-out the wavefunction of a single photon becomes, it carries with it, as itself, at every location of its being, the capacity to do a specific amount of work. The moment it "does the work," it collapses and vanishes from everywhere else it was before that moment. More accurately, at that moment, as a function and unit of energy, its capacity to "do work" instantaneously disappears everywhere equivocally although the shape of it may simply dissolve into the field continuously, no longer able to interact. Intuitively, this is hard to accept, but it is observed experimentally and forms part of the quantum's counterintuitive dynamics.

A short aside, if you cannot accept collapse, the other and most realistic interpretive option is that the photon interacts everywhere that it possibly can with each interaction resulting in a slightly different but entirely new and equally actual and very real world. Every interaction location a bifurcation of the universal wavefunction. Unknowingly, we are constantly being relegated to one of its ever-proliferating branches and although they are all thought to co-exist equally and simultaneously, communication between them is impossible. Philosophically, I do consider us to be situated in Infinite, but the *copyfication* of implied by Many-World's interpretation reveals, to my mind anyway, a reality rendered devoid of meaning. And although it may not seem absurd when thought of in terms of the Absolute, that is, that every possibility that can happen, does, or will happen, Many-World's remains too much for my philosophical tastes.

Interaction / Resolving the Conflict Between the Discrete and Continuous

Returning to Planck's quantization of energy and its many implications, the mechanism of which not only tells us of the cohesive unity and individuality of quanta, but that their creation and annihilation must be an indivisible, discontinuous, and thereby instantaneous process. This "jump" from zero to one, while skipping intermediate stages, corroborates the spookiness of its discontinuous, and immediate action. An action we also see when we observe collapse but is *implied* here by the very specificity of the constant h.

Envisioning the instantaneity of Planck's quantum of action and reconciling discontinuous quantum "jumps" with smoothly varying waves may seem an impossible task. However, quantum interactions should be seen as pieces of energy falling out of a continuous spectrum, like a drop plucked from the ocean. Quanta themselves, although highly unified individuals, are continuously varying waves that modulate the quantum foam as Aether. It is not their shape or structure that is discrete but only their method of interaction, their *energies* that are discrete, and in interacting in they way they do, quanta show themselves to be units. This gives to physical objects an induvial identity of minimal relation while also allowing them to be, each in their own way, a "part" of some larger-order whole.

Let fall two drops of water, each representing a quantum of energy, into a still pond. The circularly expanding wavefronts are each their own objects, but they are more deeply a shape of the water itself. It is in this way that we reconcile notions of the discrete with the continuous, and it is in this way that you and I are One. Except in our case as living beings, that "water" is a timeless sea of sentience that we each express as our own self.

The essence of Planck's discovery is that energy is proportional to frequency and wavelengths — being integer valued wholes — allow for its discretization. To see how continuously varying waves can embody elements of discreteness, we need only consider stationary states subject to boundary conditions. Geometrically, a boundary condition binds and defines the wave shapes a particular system can have. Take a guitar string attached at both ends; it can harbor many integer-valued wavelengths — each a "mode" of the string — each discretely different from the next. To set the string oscillating, a quantum, a minimal amount of energy is required. Then, as more quanta are added the string reaches higher and higher modes. Again, as each unit of energy moves into the system, the waves produced will be distinctly different and well-defined, but we will no longer know which quantum of energy went where, only that the total configuration is made up of a specific number of quanta.

From the previous discussion, a puzzle arises best illustrated by asking a question. Given that a quantum is defined by a period of its vibration (frequency) and possesses spatial extension (wavelength), (that is, it possesses spatiotemporal extension) how is it created in an instant? This problem led Arthur Eddington, the famous astronomer who provided the first experimental proof of Einstein's theory of gravity, to assert that the quantum doesn't "hang together" as it "seems to have no coherence in space," but rather "a unity that overleaps space."˙ With its "unity that overleaps space," Eddington is referring to its cohesion as a nonlocal object and its ability to reorient itself in an instant. Which, if we accept that these quantum waves collapse and annihilate instantly, then why not be created instantly?

Regardless, these mysteries don't have easy answers, but here's why I think quanta are countable objects with discrete energies that exhibit the spooky actions of instantaneous creation and collapse. The simple answer comes from computer science and is the reality of physical information as a binary phenomenon. A bit of information is fundamentally a dichotomy, the ultimate either/or and every quantum interaction represents the flipping of the universe's bits. As such, the capacity to carry and propagate information may be the sole "cause" of instantaneous collapse as its transference can only result in $|this\rangle$ or $|that\rangle$ outcome — one or the other with nothing in between. The transference and computation of information may be why quanta must instantly disappear from its many other possible interaction locations. It may be a simple book-keeping requirement for the computation of the universe's machine-code.

Likewise, we are also free to imagine the creation of a quantum as a "layer shift" between distinct but spatially inseparable fields. That fields spatially overlap but remain distinct may be the cause of the quantum's instantaneous disappearance from one and appearance in another. When energy moves between them, in one instant, one field is without a quantum, and in the next, it is with a quantum. The shape of all fields involved are equally distorted and remain the same as the quantum itself and doesn't move anywhere. An electron as quantum of electron matter-field energy "switches" into the adjacent, always-already there, EM field, thus instantly transforming into a photon, a quantum of radiation-field energy. A quantum, seen in this light, would be defined as a difference in field activation. In this layer shift analogy, the quantum needn't move through space or time in any sense; it merely shifts

---

˙ Eddington, Arthur.: *The Nature of the Physical World*. (1928)

or "jumps" between and activates different fields, different properties of the vacuum we call Aether.

Solidifying the Reality of the Quantum

A few years after Planck introduced his quantum of action and universal constant h, Einstein solidified it by using his "energy lumps" to explain the photo(n)electr(on)ic effect, a radiation to matter interaction. Historically, the photoelectric effect was an observation made by the same man who experimentally verified the existence of EM waves, Heinrich Hertz. Hertz noticed that when he shone a light onto a metal surface, it freed some of its surface electrons. Knowing that every metal possesses a "work" function that must be overcome, Einstein was able to show that the energy needed to eject an electron from the metal's surface was the photons frequency times h minus the work required.

Given all this, one would think that they could eject more electrons by increasing the intensity of the incident light. But light is made of photons, which are in turn a special class of as quanta called bosons. The defining feature of bosons is that when they get together, they sum. One can add (or better, superpose) equal wavelengths of light (identical energy photons) to increase the light's amplitude, which does increase its intensity. But this kind of increase in amplitude/intensity is achieved by summing equal energy photons. The metal's awaiting electron will pluck out only one of them, and therefore, if a single photon does not possess the energy to overcome the work function and free it, no number of superposed ones will either. What is required to release electrons are individually high-energy photons. So, an increase in *individual* photon energy is needed, ones that will consequently possess a shorter wavelength.

With his Nobel-winning work on the photoelectric effect, many believed that Einstein had shown that light — correctly known to be a wave — was also somehow a particle. In truth, what he illustrated was the unity of the quantum and of the necessity to take seriously the idea that they can collapse. The fact that electromagnetic energy is quantized does not necessitate turning a wave into a particle. Quanta merely deposit their energy whole and as units — all or nothing. Their unity *requires* that they interact *like* a particle. But this does not mean that they are or ever become particles. Photons are spatially extended waves capable of instantly collapsing into and depositing their energy into atoms. In the case of the photoelectric effect, it is an energy that both frees and becomes the electron it ejects.

The Instantaneous Non-Action of Entanglement

Although Einstein was wrong about quanta's particulate nature, his contribution was crucial for the conceptual distillation necessary to understanding the quantum as a natural and unified object. And even though Planck and Einstein had worked all of this out in the early nineteen-hundreds, the incongruencies and nonlocal aspects implied by the unity of the quantum went unnoticed. It wasn't until after the theory had taken on its mature form, after Heisenberg and Schrödinger crystalized it in the mid 1920's, that Einstein vividly saw the kind of spooky action that had been implied the whole time.

To Einstein, this impossible action signaled that the theory was incomplete and in a landmark paper with collaborators Boris Podolsky and Nathan Rosen, he set out, and later failed, to prove it. As it turned out, to bring to light the actual, instantaneous mechanism inherent in quantum systems required the consideration of an entangled state, the contemplation of interacting quanta. Einstein noticed, when considering the dynamics of interacting quanta, that the interaction itself would *entangle* the two. This entanglement meant that they'd come to share a single quantum state and consequently, forever thereafter, they would share an instantaneous connection regardless of their spatial separation. If one were to measure a property of one of them it would, at the very same time, cause the other reconfigure itself accordingly.

Prior to any measurement the state of the whole is indeterminate, but in measuring, say, the momentum of one *causes* the other adjust itself accordingly, instantly. Their connection is revealed as nonlocal correlations between them. There is no way, even in principle, for either quantum to know what might be asked it, but we know when one of them is posed with a question, entanglement assures us that the other will instantly "know" and answer by adjusting its state accordingly.

In other words, when two or more quanta entangle, a novel and shared quantum state develops. In essence, the previously two become one. And although this could be seen clear as day, Einstein could not accept it as their relationship can violate locality, possibly his most cherished physical principle.

Naively considered, entanglement certainly does violate Einstein's maiden. The locality principle is the common-sense notion that asserts that only objects placed next to each other can affect one another and at a speed no faster than light. It is the antithesis of action-at-a-distance and is trivially true of macroscopic objects, but information-carrying waves in quantized fields can violate this principle. But it is no "true" violation as it does not involve or permit a transference of communicable information and so it does not violate relativity's ban on superluminal signalling. Nothing is transmitted or conveyed, only a novel state adopted.

The magic of entanglement is that it permits two quanta to become one. It unifies disparate quanta, allowing them to then "overleap space" and behave as a single quantum would. The 'previously two' are now one and will remain instantly connected to and may affect one another across any distance.

A New Era

The discovery of the quantum and introduction of Planck's constant brought instantaneity and discontinuity into physical theory and ushered in a novel era. Einstein went on to use Planck's quantum of action to resolve the photoelectric effect, whose mechanism of collapse falsely projected radiation as having a particulate nature. It doesn't, it just interacts like it does. Quanta are not indestructible; they are fickle and fragile, easily disturbed, and affected. They are ghostly apparitions, created instantly, whole, and unified, and because of this, they are countable. That they can be created in an instant also implies that they annihilate in an instant. Quanta either behave quite strangely, and their entangled states harbor instantaneous correlations, or the world is an endless bifurcation of branching worlds, making copies of copies, of copies, of copies, of copies... *ad indefinitum.*

After Planck and Einstein had solidified the constant h, Niels Bohr was next to put it to use. In 1915, the common conception of the atom was that it mimicked the solar system. Ernest Rutherford had proven that atoms had tiny, marble-like cores — held together by forces far more significant than electromagnetism could account for — that was thought to be surrounded by point-like orbiting electrons. A beautiful picture, but this planetary Rutherford model of the atom had deep flaws. If the electron truly were a point particle orbiting the nucleus, it would radiate EM energy and spiral into the nucleus in a picosecond. Using Einstein's energy "lumps," Bohr was able to explain the atom's stability. Essentially, Bohr quantized the possible orbits of the electron into different energy levels, with one as a lower limit. But to explain why they needed to be quantized was left to Louis de Broglie, for he was the first to envision the electron as a wave of matter.

Nevertheless, Bohr's contribution led to QM's first success; an explanation of atomic spectra, an element's bar-code-like signature written in light. Every elemental atom produces a unique light spectrum based on its electro-nuclear structure. As such, atoms with different amounts of nucleons (elements) are known to capture equal numbers of electrons around them. When these electrons change states, they are responsible for the production of predictable frequencies of light. As they gain or lose energy, they "jump" from one orbit to another, creating a wave of electromagnetic radiation. The wavelengths of this radiation can be measured and tell us exactly which elemental atom produced the wave.

Today, the fields of astrophysics and astrobiology utilize this technique to study the atmospheres of exoplanets. Astrobiologists look for markers that may signal that life has taken hold on other planets. Using this technique, astrobiologists recently discovered phosphine gas on Venus. A gas which, on earth, owes its existence to living processes. This discovery hints toward the possibility that life, in whatever form, may have a foothold in the noxious atmosphere of the planet.

Following Bohr, Louis de Broglie introduced the idea that the electron may be better understood as a wave. Could there even exist such a thing as a wave of matter? By postulating a wave nature inherent to the electron, De Broglie explained Bohr's *ad hoc* orbits. He illustrated that the atoms' stability could be intuitively understood by envisioning the electron surrounding the nucleus as a wave of matter, a fluid-like oscillation of electric energy bound to the core. Rather than the planetary view of a point-particle circling the atom's center, through de Broglie's efforts, it was realized that electron orbital configurations were various shapes of 3-dimensional standing waves, with the nucleus serving as its boundary condition.

Finally, enter Erwin Schrödinger. Building from De Broglie, Schrödinger was able to codify his insight into a rigorous mathematical framework. A wave equation that bears his name forms the heart of modern quantum physics. Solutions to the equation reveal that the electron is a cloud-like pulsation of electric field energy captured by the nucleus. The gorgeous shapes of which are 'somewhat' revealed by it because quantum waves are complex in the technical mathematical sense possessing both "real" and "imaginary" parts. Thus, the shapes shown by Schrödinger's equation portray only a sliver of a snapshot of the electron surrounding an atom.

Furthermore, because the electron is a matter-wave captured by the nucleus, it also acts as a well of electromagnetic influence, shielding the nucleus from radiation. Seen in this light, an atom-captured electron can be thought of as a warping of the EM field, just as in general relativity localized aggregates of energy-momentum warp the cloth of spacetime into a well of gravitation.

To recap some of the main insights brought about by our semi-historical exegesis. As Planck uncovered but did not realize, quanta's all-or-nothing character speaks to *the quantum of action's indivisibility and instantaneity.* As such, quanta collapse whole or not-at-all and always instantaneously, regardless of spatial extension. This mechanism is responsible for the discrete "quantum jumps" or "teleportation" style of movement they are counterintuitively known for. To be fair, a quantum jump may take some amount of time, no matter how small. Regardless, it is undoubtedly so fast, indeed, experimentally verified to be superluminal, that it appears to us as instantaneous. Possibly, refined future experiments may place a duration on "jumps" and "collapse." For now, we are confident they take place as near as an instant as is measurable. This superluminous aspect of quanta that allows them to "overleap space" does not violate relativity as information is not communicated by this mechanism.

Bound to the atom, an electron (itself a quantum) can "jump" from one orbit to another without visiting the space in between. The quantum of action is the discrete process when an electron dumps some of its energy into the EM field, a simple transformation of energy that creates a photon. The energy always remains just that, energy. The vibrations of one field put in motion the vibrations of another, and curiously, at the location of the interaction, never "go anywhere" but merely moves between fields. The quantum of action is also present when we observe quantum "collapse" in experiments. The whole extended quantum acting as a single individual goes from being "all-over-the-place" to just a single interaction location in (possibly) zero-time. As we have seen, Planck discovered the quantum of action as a proportionality constant between the energy and frequency of an electromagnetic wave, but its usefulness has far surpassed even the simple matter-to-light interaction.

$$\hbar$$

What of information? How does a quantum harbor and communicate it? In the form of qubits of course. A quantum's "spin" superposition is the most elemental qubit. Like a bit, it is always an either/or "object," but being a quantum is either/or at the same time. Much more will be said of spin later but know that spin isn't the only qubit contained in a quantum. Concerning the information-theoretic machine-code layer of the vacuum, every quantum carries with it many qubits of information, each specifying a particular property or degree of freedom inherent to it. More than either/or, qubits can exhibit a superposed spectrum of ratioed values concerning, for example, its momentum or energy.

Every time quanta interact; they exchange information by performing natural logic operations that transform and flip qubits and cause them to adopt actual values before evolving into superpositions again. This is but another way the universe "keeps track" of its evolution. As we saw with the geometrically bound standing waves, the fact that quanta are waves leads directly to the quantization of these parameters. That is, how a wave can be both a continuous object and embody discrete elements.

Quanta are countable because of the information they contain and it is their unity that allows for "spooky" action at a distance. The fact that quanta act as units despite their physical continuity is a discrete property. A feature that may be indicative of the binary nature of the information it carries.

On the immaterial, albeit substantial, quantum layer, a quantum is an idealized excitation of a field, when created truly — with enough energy — it becomes a field's *resonance*. As we will see as we proceed, the mere movement of these waves results in self-interactions that cause various *disturbances* in adjacent fields that lead to the more physically accurate concept of a wave packet rather than a simple, lone wave. Quanta, in fact, all physical objects, are never truly isolated. Anything and everything is locally and causally in touch with its environment.

An electron, for instance, is a spread-out energy and momentum *density*, a conglomeration of dynamically interacting waves in a few fields that include psi, Higgs, and electromagnetic. The field for the *ideal* electron is the electron-positron field, sometimes called the Dirac field or psi, always denoted by the Greek letter Ψ. Again, the discrete nature of these waves and the information they carry lead to the fields being quantized. That is, these waves are individuals and when they interact, they "collapse" instantaneously over large distances. The process of collapse will be considered when we examine decoherence. But for now, let's consider some of quantum theory's more fantastic phenomena.

"*The essential reality is a set of fields* subject to the rules of special relativity and quantum mechanics." - Steven Weinberg[*]

CHAPTER XI

Quantum Waves | A Torus of Tides

A quantum is a warping resonance of a field, characterized as a wave; it is a vibration or movement of some medium caused by an energy that it itself *is*. The wave carries information and moves through the field as an energy-momentum-charge density and can be said to *be* those quantities.

We may feel as though we are going backward but lets for the moment consider a very basic question: what exactly is a wave? A dynamic shape of its host medium? Is it more than energy and information? And how do they shift phase or spin?

Some have it that a wave is not a thing, not a noun, but is instead a verb, a something that something else does. And although this is partially true, it is incomplete as every wave has its own locatable, historical genesis and this moment confers upon it a kind of originality and individuality. Given that, can a wave be considered an object, its own entity?

Let's create one and see. Secured to a wall, we begin by holding taut a rope and shaking it up and down a few times. The action creates a wave with as many crests and troughs as the shakes put into it. The created wave will move along the rope's length and reflect when it transfers its energy into the wall. Due to the elasticity and internal friction in the rope, the wave's energy will dissipate. The rope's elasticity dictates the ease with which the wave can move through it and as a consequence will define its speed. The higher the elasticity the quicker the wave will lose its energy. Also, in this example, our wave will lose energy by fighting gravity and internal friction when bending the rope.

In this simple example, we see how a ripple in a medium can move energy and information. It took a certain amount of energy to shake the rope, and the number of shakes is information as a record of frequency. But the wave is more than energy and information as it's also a dynamic shape of the rope. The wave is a thing, indeed its own object, even though it relies on factors beyond its control to come into being.

---

[*] Weinberg, S.: *The Search for Unity.* MIT Press (1977)

Also necessary for a wave to obtain is a medium and minimal amount of energy. In our example, the rope served as the medium, but we saw that it is of such a nature that the wave will quickly die out. Having had its energy bled off by the rope's elasticity, gravity, and friction. A quantum field, however, is no mere medium but a property of space. As properties of space, quantum fields and their waves can exhibit behaviors impossible in classical physics. It is possible that a true resonation of a particular field may oscillate indefinitely. And the elasticity of massless fields allows for light-speed travel, in fact, requires it. The unity of wavelike quanta allows them to "overleap" space. But how are these waves captured and described in the formalism?

As we know, the wavefunction mathematically describes all the information quanta contain, and how they behave, interact, and propagate in their fields. It is a mathematical description mirroring a physically real quantum state, and it characterizes the kinematic evolution of a complex wave in a field. The wavefunction's progenitor, Erwin Schrödinger believed that it represents an energy-momentum-charge density of a quantum field.

By squaring the wavefunction to remove negative values, we arrive at a probability amplitude for an interaction location where the quantum wave *may* collapse; a place where quanta interact, reduce their states, flip their qubits, and transfer their energies into adjacent — albeit spatially inseparable — fields. The "squaring of the wavefunction" is the Born Rule, whose interaction location is interpreted as the manifestation of a particle but is simply the place where the wave collapses, vanishing into another field. One quantum annihilates while simul-instantaneously creating another.

In the equation that captures the dynamics of wavefunctions, we encounter the mystical number i. A number that signals us to look for wavelike, oscillatory phenomena. A *complex* number that possesses both real and imaginary parts; it is the square root of a negative one. Its 'imaginary' part is a bit of a misnomer because both real and imaginary numbers exist in the same way. Ontologically speaking, they have the same abstract status as they are both equally mathematical objects and are as real as numbers get (which is *real* by the way; numbers, despite their timeless, aspatial mode of existence, actually form aspects of the Infinite).

In QFT, complex numbers are used to assign to a single point two dimensions or values. These are needed to describe waves as they have both an amplitude and a phase. Due to this and a few other facts, it's impossible to form a complete mental picture of a quantum object. There is simply too much going on; the degree of complexity is simply beyond our representational capacities. Even the shift from visualizing a 2-dimensional wave on the surface of a pond to a 3-dimensional one spherically expanding from a radio antenna is hard enough. Now, add even further dimensions in which they can oscillate, both real and imaginary, and grant them a helical polarization, twist that twist again and you start to see that you cannot "picture" a quantum. Regardless, we know they possess certain properties, elicit specific characteristics, and behave in statistically predictable ways, so we can form a model, albeit an imperfect one. To arrive at it, let us consider the properties of waves.

A wave's **amplitude** is its deviation from the zero value of its field. The peak of which is called its crest and the base its trough. Consider an ocean wave as it moves past an anchored buoy in one dimension. As it approaches, the buoy begins to rise until it peaks and begins its descent into the trough, finally returning to its original position. Another different internal degree of freedom belonging to matter quanta — spin — can also cause the same deviation in amplitude but it achieves this by giving the wave a spiraling, involutive, helical aspect. We'll examine spin in greater detail later but through our brief examination of amplitude, we may tease a few insights; we see that the water is only there "to allow" the wave to be. That the motion of the medium differs from the movement of the wave. We see that the wave is its own object with a well-defined structure. The maximum amplitude of both its positive and negative values is an equidistant deviation from zero in opposite directions. We can also see that a wave possesses an interrelated **frequency** and **wavelength**, such that the two define each other. The wavelength is the distance between crests, and the frequency is a measure of how quickly the wave moves through a point in time or, following our example, the buoy in the water.

Wave **phase** is the linchpin of quantum theory. Without it, none of the interference patterns that emerge from the experiments would make any sense, and the probabilities for interaction locations would not even exist.

A wave's phase is its shape and "location" of where it is in its propagation. Of any repetitive waveform, phase specifies the location of a point within its cycle, a place on the shape of the wave. This concept is captured mathematically by degrees or radians. When two or more waves meet, their relative phases dictate how they will sum and what kind of interference pattern will emerge.

To understand phase somewhat more intuitively, consider the moon also has a 'phase' with which we will compare with that of a wave. At zero degrees, i.e., zero phase, we have a new moon. As it waxes crescent and moves into a half-moon (90 degrees), this would be the crest of our wave, also its peak amplitude. Waxing gibbous to become a full moon would represent the wave's half-wavelength node (180 degrees). In a regularly repeating standing wave, a node is a point where the amplitude is unchanging and, in this example, this wave's node is halfway through its cycle, or period. Each stage as it wanes back to a new moon is the trough. A wave's phase is also its whole configuration; the how it is, where it is, and what shape it has at a specific place in space and time.

The wavefunction of a quantum assigns a field strength to its phase and amplitude. If the phase undergoes an operation, such as a rotation, the field's value must reflect that. The process by which this is accomplished is through that previously mentioned word-monster *non-abelian phase symmetry*. A "local" symmetry related to the phase of the wavefunction, it is precisely this notion that leads to "force" in QFT. Beholden to this symmetry, QFT replaces the concept of force with an intermediate field that facilitates interactions of material quanta. The actuality of this 'interaction' field is *required* by the symmetry; to register a quantum's phase "shifts," and this ability to shift and feel the shifts of other quanta is more commonly known as **charge**.

A **phase shift** is when a quantum's phase "jumps" from one position to another. In our moon analogy from before, it's akin to the moon suddenly jumping from a crescent to half-moon, skipping the intermediate steps. Again, in the formalism, the description of a wave involves pi (π) as its phase is measured in degrees or radians. One can model or more easily see a phase shift as a skip in degrees around a circle as if the phase suddenly jumped from 45 to 90 degrees. If an electron's phase shifts, due to the necessity of the conservation of charge and their symmetries, another field must register and propagate that change away. This necessity of requiring local symmetry to propagate away change is the origin of force fields. As such, an electron's phase shift creates a virtual photon that can be absorbed by another electron or reabsorbed by itself. When absorbed by the other, the result is a "force" felt between them. A binary function, depending on whether the phase shifts are clockwise or counter-clockwise dictates whether the force that manifests repels or attracts.

Charge is a conserved quantity that relates how two objects in the same field interact through an intermediate field. By electrostatic charge — what left Thales dumbstruck — we now understand as the electric field itself. By shifting phase, electrons create "virtual" photon fields around them. Again, the direction of the phase shifts will dictate whether they will register as positive or negative charge and will attract or repel one another accordingly. Electrons never touch but interact by passing virtual photons back and forth in the EM field.

Let's look a little deeper; an electron possesses — and therefore is — one unit of negative charge but this "unit of charge" is more accurately a "virtual" photon cloud produced by the electron's constantly shifting phase. Said differently, "charge" and "phase shift" are synonymous terms. Furthermore, because all electrons phase shift in the same way, they will always repel one another, unless of course the are coerced into an orbit by the weight of few protons. Which, if a proton it to *attract* an electron, it must, as a composite whole, shift its phase opposite to that of the electron. We can see here another consequence of entanglement. That countless quanta will always sum to one.

Energy and Momentum

How is it that a wave can possess properties such as energy and momentum? Special relativity tells us that energy and momentum can be considered two facets of the same phenomenon seen from different frames of reference. Similarly, frequency and wavelength share the same fate. As Nature would have it, the frequency of a wave is proportional to its energy while its wavelength is proportional to its momentum.

On energy, we know from Planck's work that the energy of a wave is proportional to his constant h times the frequency, an integer-valued quantity. Consider the standing wave of a vibrating guitar string, only discrete modes or values of wavelength are allowed. As we add more wavelengths to the string, they get shorter. Keeping its structure imagine turning it into a traveling wave and seeing its frequency.

To see this intuitively, let's return to our rope tied to a wall. One must pull up and down on the rope to set a wave in motion. If one pulls up too slowly nothing will happen. Due to the elasticity of the rope, there must be a certain amount of effort required to get a wave moving, a certain amount of energy. There is a critical threshold that must be met to create a wave. This threshold is determined by the elasticity of the rope, the wave's medium. If the minimum requirement of energy is met a wave with a long wavelength will be created. Adding more energy by moving the rope quicker results in a shorter wavelength and thus higher frequency. We can see here an intrinsic relation between wavelength, frequency, and energy. Higher energy is proportional to a higher frequency and correspondingly shorter wavelength.

Of momentum, we can understand intuitively how it relates to the wavelength by considering what happens to a wave's slope when more energy is applied. A wave's momentum is exploited by surfers. If a wave's slope isn't steep enough, it won't be able to transfer its momentum to the surfer and she will not be able to 'catch' the wave. Waves with long wavelengths and thus shallow slopes are unsurfable. But a wave with a shorter wavelength and thus steeper slope will allow a surfer to ride it. One might object that it's gravity that pulls the surfer down the slope, but they would only do so by forgetting the wave is constantly pushing them up transferring to them the momentum that is in the wave itself. Thus, we see that the momentum of a wave is inversely proportional to its wavelength and/or proportional to its frequency. The fact that waves are surfable illustrates the fact that waves possess both energy and momentum. Compressing a wavelength leads to higher momentum while increasing its frequency leads to more energy.

Resonance / Real and Virtual Quanta

We have, in passing, briefly met with "virtual" quanta. These quanta are understood as "distortions" in the wavefunction, the pieces of which do not possess enough energy to become a field's true resonance. Virtual quanta are said to be off "mass-shell" and so can *momentarily* break some rules. Due to the conservation of total energy, when an electron shifts phase, this difference must be "propagated away" by another field. So, an electron's phase shift creates a "disturbance" in the EM field, but not a true photon, the EM field's true resonance. If every phase shift were to produce "real" photons, the electron would quickly evaporate into radiation. For this reason, an electron primarily creates only disturbances in the EM field, thereby cloaking itself in a haze of virtual photons that it continuously and endlessly reabsorbs.

A virtual photon cloud is a kind of electron-cloaking fuzz that the electron is incessantly producing and "hiding under." The attractive or repulsive force that manifests between electrons and positrons is through an exchange of these virtual photons. It's as if an electron is continuously throwing out virtual photons but reabsorbing them before they have gone so far as to be lost.

Likewise, but more complicated, in quantum chromodynamics, a quark's phase shifts create the glue that binds them together, gluons. These gluons are then felt by and reabsorbed by other quarks *and* gluons in the maelstrom marble hearts of matter. When a "red" quark turns (or phase shifts) into a "blue" quark, it creates a gluon that carries the difference away, to be reabsorbed by a "green" quark a picosecond later. The phase shifts in QCD are more commonly known as color charge.

Now, a quantum is an amorphous, energy-momentum density of a few fields, better characterized as a wave-packet; a conglomeration of waves swimming in a few fields. Take the ideal electron, a *real*, energy-momentum-charge density of the Ψ field. It's a flowing movement of spatially extended energy with the characteristics of a wave that carries information. This wave is always-already flowing in the space of — and coupled to — other fields and behaves according to the potentials it feels. One of which, due to its spin characteristics, is the omnipresent Higgs field. Some of the electron's energy must go into it and so is slowed down by — and (as we shall see shortly) gains

Quantum Waves / A Torus of Tides

its small mass from — its constant coupling to it. Thus, the electron is always-already coupled to and causes oscillations in the Higgs field.

The Higgs is a scalar field, and its value moves up and down and is pushed and pulled by an electron as it passes through. As such, these two fields — electrostatic and Higgs — work in tandem to become an electron. But these are not the only fields involved in forming the wholeness of an electron. For an electron to feel and respond to other electrons it must also surround itself with a cloud of *virtual* photons. Thus, it possesses electromagnetic properties and responds to the potentials of the photon field.

Only by its virtual photon cloud can an electron "feel" other electrons and positrons and adjust its trajectory and shape accordingly. Two electrons never touch but feel one another and communicate their intentions by *disturbing* — but not truly setting up a resonance — in the EM field. If the electron set up a true EM field resonance — created a real photon — it would lose that energy and quickly evaporate into radioactive dust. It's by this "exchange of virtual photons" that the "force" is generated between Ψ quanta.

But what is really going on here? What is a "virtual" photon anyway and how or why does an electron cloak itself under them? Furthermore, what is the difference between a real and virtual quantum?

Real quanta are (almost) timeless, true *resonances* of their fields. A resonance is a proper oscillation of the field, whereas a disturbance is an incomplete and unstructured vibration.

Remember, in an atom, an electron can drop an energy level and create a photon of equal energy. These will be entangled, and this entanglement will be affected by future interactions, but each will maintain a semblance of their own identities and wholeness until said interaction occurs. This is what we expect due to the conservation of energy. As we shall see, due to entanglement, they will also carry away information about the other and remain instantaneously correlated. They will come to share a common qubit and will each be considered a real quantum. A real quantum is said to have an energy-dispersion relation that is on "mass-shell." Put simply, this means that the quantum has enough energy and internal coherence to live an extremely long life.

A field is like a tuning fork that will only entertain notes beholden to itself. Consider, a G-tuned tuning fork will only sound off with the resonant frequency begetting of a G-note. The tuning fork will not ring out with any other tone. Likewise, a field's true resonance is like the G to the "G-field." In the electrostatic field, an electron is the field's true resonance. It is the "sound" a field would echo if it could make a noise.

If a quantum does not possess enough energy to put into the electron field and set up a resonance proper, it will not resonate in its field but will instead dissolve into radiation very quickly. The kind of quantum that is unable to become a "real" quantum is said to be a "virtual" quantum. Virtual quanta exist and momentarily possess the characteristics of their real counterparts, but their short-lived lives do not satisfy a proper energy dispersion relation and as such, they vanish very quickly.

Now, what about a free electron; one released from its atom? How does that look as it propagates in its field? As we know, all alone, it behaves rather complexly for as an electron travels it shifts phase. That is, a point in its phase will skip a beat, jumping from — just as an example — 45 to 90 degrees. Due to symmetry, this shift must be "carried away" by a newly created quantum and this is where the virtual photons come in.

When an electron shifts phase, it must place *some* of its energy into the EM field to excite it and in so doing creates a virtual photon. It's a virtual photon because it does not possess enough energy to become a true resonance of the EM field and so it will "die out" or almost instantly be reabsorbed by the electron. If the electron gave enough energy to the EM field to create a true resonant photon, the electron itself would quickly evaporate into EM radiation. To complicate matters, by "losing" some of its energy to the EM field, when an electron shifts phase, it itself becomes a disturbance, a virtual quantum of the psi field.

Virtual quanta are so fleeting, so ephemeral and vaporous that they scarcely exist and as such can operate underneath Heisenberg's indeterminacy principle. To come into being, they can "borrow" energy from the vacuum, so long as they return it before anyone notices.

So, what is an electron "really?"

To attempt an answer, an electron is a spread-out (or spreading out, depending on if it's free or not) energy-momentum density of the coupled psi, electromagnetic, and Higgs fields; a spinning and constantly shifting phase wave that surrounds itself in its charge density, a cloud of virtual photons. Another name for this picture of an electron is that of a "dressed" electron, you can guess what the ideal electron is called.

Combining the vibrations of these fields, this composite structure can be considered a single ripple in a universal matter field. Despite the many fields involved in making up its structure, it forms a composite whole that we understand simply as an electron. Therefore, the unifying expression of three distinct fields nevertheless forms a resonance in a higher-order field. We have here again that superposed mereological process where many fields gather and form a single field. Just as many quanta can gather to form a unified quantum of larger size, complexity, energy, and information content.

Now we can pose the question: do these fields, when coupled together and becoming another, generate and form a novel field or does their instantiation together access a deeper, always-already there, just not 'activated' yet field? I believe it's the second aspect, and that the field of consciousness is similar. Always-already there, but not yet deviating from its lowest energy state because the unified quanta have not yet achieved that certain level of layered complexity.

Regardless, we should note, that one never really gets to the bare electron, for in all our experiments we only ever "feel" or interact with the fuzzy virtual photon cloud that surrounds it, the electrons suit of EM armor, if you will.

Becoming Matter

The furniture of reality is formed of a single material substance: matter. The three primary properties of which are mass, spin, and charge. If these can be known, all others can be derived. Mass, however, holds the primary position as its emergence marks matter's primeval property. The desire is to know how Nature achieves "solidity" and *mass* is just the term to get us there. But deep down, matter is truly, and nothing other than, structured energy. So how is it that we move from immateriality to materiality? In other words, how does a "baseless fabric" beget reality's basic building block? The short answer: by having energy quanta respond to and distort a particular *set* of vacuum fields.

We have seen how a wave can possess the classical variables of energy and momentum, but we have not yet seen how a wave can become "massive."

Just how exactly does an immaterial energy wave acquire inertia — the property of mass? Before considering the long answer to that, let's first make a distinction between free and captive quanta. A free quantum can evolve in its field unincumbered, that is, on its own without external influence. While a captive quantum is one irrevocably involved in an endless exchange of virtual quanta with one of its brethren. In the forthcoming section *the music of the spheres,* we shall see how captive quanta gather and entangle to form matter and atoms as we know them. Free quanta are those that we liberate from their couplings to participate in our experiments. They are those not bound to others. That said, how does a complex energy wave acquire mass?

In two ways, the first and most fundamental is by coupling a quantum-as-wave to another field thus forcing it "feel" another potential. Even a free quantum — that is, not bound to an atom or molecular system — can acquire mass in this regard. The second way and more important for mass as we know it is by a principle known as confinement, a process wherein at least two (or more) material quanta get together, hold one another captive, and through their interaction dynamics of geometrically confine themselves into a stable system (the atom) and bring about "solidity." Confinement is the core element of quantum chromodynamics, the theory that explains how we acquire the lions share of our mass, but again, the deeper reason is the mere coupling of fields as even confined, chromodynamic quanta require this as a first principle. As such, we will examine field coupling first and confinement when we visit quantum chromodynamics.

Imagine a photon, a wave of energy racing through the EM field. As we know, it always travels at C, due to the field's base-level elasticity. It is this "cohesive character" of the field-as-medium that determines the velocity of the waves that propagate through and within it, as well as defines the relationship between a wave's frequency and wavelength.

The easier for a wave to move through a particular field, the higher its elasticity; it is the ease by which the traveling wavefront "pulls up," thereby activating, the next part of the field. The lower a fields elasticity the more difficult it will be to do so and consequently; the more energy will be required to displace and thereby activate it. Consider, fluids of different viscosities have different elasticities. It is far easier to set up a wave in water than oil or tar and their wavelengths will reflect that. This relative 'easiness' of wave production is reflective of the fact that it takes different amounts of energy to create waves in different mediums and demonstrates the bare fact that energy is required to 'disturb' the field in the first place, even if that field is massless.

Now, what if we could change the properties of a field's elasticity? What would happen? In other words, what if a quantum were required to adhere to a new potential, by overlaying it with another field, and forcing it to couple to it on the spot? The answer: mass. But not yet mass as we know it. For at this level, we have only achieved a kind of vaporous inertia. A wave has moved from oscillating in a massless ocean of air to fluctuating in a sea of nearly substance-less substances.

Because they "spin" in a complicated fashion, this is exactly how fermions — matter quanta — attain their first foray of substantialization. All solid matter — that is, all the "hard stuff" — in the universe is built up from just three simple fermions; up and down quarks form the nuclear heart of atoms, and electrons form the electrochemical clouds that surround them.

Quarks and electrons are the two types of elementary quanta that achieve initial inertia by involving the Higgs field in their constituting structure. Both quarks and electrons belong to their own fields, have different "spin characteristics," and their phase-shifts result in different kinds of charge. That is, their phase-shifts produce virtual quanta in a necessary interaction field and so interact with via their particular brand of interaction-carrying bosons.

Regarding resistance to acceleration or simply mass, it is by that always scare quote surrounded term "spin" that certain quanta couple to the Higgs, and by this union Nature arrives at Mass Without Mass; of how a "baseless fabric" becomes a rudimentary base. A note on spin, it is usually found in scare quotes because it is not at all apparent how a quantum as wave realizes it. Nevertheless, we shall make a concerted effort to intuit it in a future section.

As we know, all fields are always-already present to one another because they all equally occupy all space. Therefore, it is the nature of the quanta that moves through them that will define which fields need necessarily be involved in their constitution. Quanta that "spin" do so in such a way that the potential provided by the omnipresent Higgs field must be considered and represented in their constitution. *Some* of the energy that a quantum imbibes must go into the Higgs to displace it and it is just this *necessary coupling* that slows them down and causes them "resist acceleration," to take on and manifest "inertia" as a novel property and feature of the cosmos.

Again, some of a quark or electron's energy must be diverted into the Higgs field to displace it and the consequence is that the newly coupled quantum feels a different elasticity and will therefore travel at a reduced rate, have a different dispersion relation, and finally, take on the property of inertia — a wave belonging an insubstantial ocean will acquire its first hint of substantiality and arrive at a very minor modicum of mass.

To reiterate, when a quantum's spin property causes it to couple together and simultaneously activate two (its elemental host field and the Higgs) or more fields, it changes the elasticity of the vacuum "felt" by that wave. The energy of said quantum system is redirected and distributed into displacing both, slowing it down, causing initial inertia, and gaining mass. This is how electrons and quarks — the only two types of material quanta that matter for matter — acquire their intrinsic mass.

However, the way by which we arrive at the solidity that characterizes macroscopic reality is through the confining feature of quantum chromodynamics and Pauli's famous **exclusion principle**. This principle forms the literal steppingstone to mass proper as it does not allow any material quantum to occupy the same state as another. By this, electrons begin to form and fill "shells" around atoms, and only by this do we begin to solidify inertia to become *matter*.

But before we arrive at this, we must familiarize ourselves with some of the quantum's more counterintuitive capabilities and behaviors.

As a precursor of what's to come, despite their composition — that is, when a single quantum is understood as involving a multiplicity of fields — unified quantum systems form distinctive individuals that become resonant ripples in novel, larger-order, and unique, *composite* fields. Owing to superposition and the indeterminate state of the vacuum, these composite fields can be considered *fundamental fields* in themselves.

Here's what I mean: when quanta of quark fields (quarks) maximally entangle with quanta of color gluon fields (gluons), the result is a unified quantum of energy belonging to the now-composite *proton field*. So established, a proton will behave just as a quantum of any elementary field would. Like an electron, a proton will "spin," shift phase, and if sent through the proverbial double-slit experiment will — just as any unified and free quantum would — evolve to take both paths. This is because the proton — a maximally entangled object — is but a resonant disturbance (a quantum of energy) in a proton field. Likewise, a hydrogen atom becomes a quantum of energy in the hydrogen field, a molecular "buckyball"˙ becomes a quantum of energy of a "buckyball" field, a virus becomes a quantum of energy of its equally complex virus field.

If this at first seems implausible or even to be a contradiction in terms, know that the void-vacuum — the quantum foam as carrier — contains within it an indefinite amount of quantum fields. When certain configurations of these are caused to operate in tandem, the result is an emergent — *yet somehow still fundamental* — field. This is because the quanta that come to operate "atop" of them share in the self-same relation to them. Mainly, they are them. Quantum systems as waves are dynamic shapes of their own host mediums. They are "nonlocal" objects and possess both temporal and spatial extension that, by Heisenberg's principle, are always-already a certain minimal extent and "size."

Later, in just this fashion, we will model the human being as a unified resonance in the endlessly creative, timeless, multidimensional mindfield. But before we do, we need to familiarize ourselves with some of the quantum's more counterintuitive capabilities and characteristics. In so doing, we will see that quanta are of such a nature that adding *any amount* of them together will always, and only ever, *sum to one*.

˙ Named after Buckminster Fuller; buckminsterfullerene — or a "buckyball" as it more commonly called — is a fullerene molecule with sixty carbon atoms bonded in such a way that it forms a shape resembling the architect's famous geodesic domes.

"Mind is a *singulare tantum*. The overall number of minds is just one. I venture to call it indestructible since it has a peculiar timetable namely mind is only always *now*."[*]
- Erwin Schrödinger

CHAPTER XII

Summing to One - I | Superposition

Superposition is a multifaceted phenomenon best understood when considering waves and the mediums that support them. A single wave is already a nonlocal object as it is spread-out and "occupies" many places at once. In terms of position states, it is always-already in a superposition of its multiply realized, and realizable, possibilities. Not restricted to its namesake — position — superposition can also refer to the summation process unique to waves. Waves "superpose" when two or more overlap in space and time — reinforcing or negating one another — to create a more complex waveform.

If one tries to think of superposition in terms of particles — a particle is both here and there simultaneously — they are led into nonsense. The illogical notion of a dimensionless point-particle being in two places at once leads to profound paradox and intuitive frustration. If we try to hold onto particles, we shall be left quoting Feynman: "I think I can safely say that nobody understands quantum mechanics."[†] But if one understands that particles do not exist, but rather a single wave is already in a superposition extended over a particular spatial area, we arrive at a simple intuition and find that we can indeed understand quantum *physics*. Remember, it's not a 'mechanics.'

Although superposition is not necessarily a quantum phenomenon, it is one of their capacities that allows them to behave extraordinarily. When thinking of superposition in terms of a single quantum-as-unitary-system, it is more common to consider its *states.* The superposition principle holds that if a quantum can be in |this> or |that> state, it can be in both simultaneously. But if one is after ease in intuition, consider the superposition principle as a

---

[*] Schrödinger, E. R. J. A.: *Mind and Matter.* Cambridge. (1958). That the "overall number of minds is just one" is indicative of our shared ground floor, or the "That" which I am calling the mindfield.
[†] Feynman, R. P.: *The Character of Physical Law.* Cambridge (1967)

wave's spatially extended yet unified coherence; its cohesivity the "overleaps" space.

Consider the waveform created by the exchange of energy when a drop of water falls into a still pond. For completeness, equate the drop of water with a quantum — a piece of energy — as this will allow us to correlate the entirety of the wave thus created with the single quantum. When the drop strikes the water, a *single* circularly expanding wave will propagate out from the point of impact. It will have a few crests and troughs, its largest amplitude in the center of these, and will be expanding circularly, *in every possible direction*, as well as dissipating. It dissipates because the water is a *mechanical* medium, and the wave loses energy to molecular friction and the production of heat.

Quantum fields, however, are not mechanical; they are bare qualities of the vacuum such that their waves do not necessarily dissipate, although they can decay into ripples of other fields. Back to the pond, at any moment — t1, t2, t3 — the water is in a particular state, the wave in a superposition. Quantum theory tells us to think of this wave as a unified object moving in multiple directions and to also note that it is in many places at once.

Now, let fall two drops of water into the pond. Each drop and its consequent wave are a state of the water, but also *both together are one state of the water*. The two combine into one, another equally valid and real superposition. Even before the waves begin to interact, one can see the state of the water, as a whole, is in a single state. A superposition state built of two waves but a single state nonetheless.

Now, when the two waves begin to interact, they will *superpose*, affect one another, and create a more complex pattern. After they have left the place of interaction the waves will naturally carry away part of the other, "rolling" in a more complicated fashion. This means our waves have *entangled* as each now exhibits the affectation of the other and so by, each carries off information about the other. Indeed, by this merger, the two become one. By observing the dynamics of waves in water we arrive at an intuitive understanding of both superposition and entanglement.

A quantum "is really" a blob of field activation. Imagine a photon created by and exiting Alpha Centauri, as it expands spatially and its wavelength is stretched by dark energy's ever-increasing expansion of the universe, while always carrying with it the capacity to deposit its quantum of energy at everywhere it is. The instantaneous creation of a quantum makes it appear as if a tiny Big Bang-type inflation takes place and the interaction location because it superluminally inflates to the lower limit of Heisenberg's position/momentum indeterminacy relation and thereafter evolves as predicted by the formalism.

To drive the point home, as will be seen by the Mach-Zehnder interferometer experiment, a single quantum's wave can be *redirected* such that it interacts and interferes *with itself* to create that more complicated pattern. By bringing two or more waves together, we are brought to the more common conception of superposition; how waves of a common medium interact when they begin to occupy the same location in space and time.

When two or more waves approach one another, they begin to overlap, and their individual phases superpose. When this happens, the magnitudes of their displacements will sum to become a single, more complex resultant wave. The resultant wave will be a slightly more complicated dynamic shape of the field, created by the relative phases of all the waves that went into it. Thus, any resultant wave will harbor the contributions of those that went into it, but one will no longer be able to identify which quantum was which. In theory, through a process known as Fourier analysis, one could tease apart any complex waveform and see what wave shapes built it up, but this will not allow us to *identify* them.

Consider identical, sinusoidal waves approaching one another on a rope. With the rope sustaining them, they cannot move out of one another's way and so must interact somehow and sum. When two waves overlap in space and their crests and troughs align, they are 'in phase' and their amplitudes will sum, the resultant wave created will have a greater deviation from the zero value of the field. To revisit the concept of laser but this time guided b the light of superposition — we "amplify" light by carefully stimulating the emission of light itself. To increase the intensity of a laser, photons must be added to it.

These are not particles but discrete increments of EM field energy, such that when a "photon" is added to a laser, what is being added is a synchronized, in-phase, equidistant wavelength of discrete energy into the field to that location. The combination of which deviates the field so strongly that the concentrated light emitted can burn whatever it is directed toward. Basically, adding energy increases the "pressure" of the photon "fluid" — the EM field activation — that forms the laser. This is all achieved by *superposing* quanta of the same state, and only bosons can do this. When waves are 'out of phase,' crest and trough align, and the field value will be near zero.

In one respect, superposition is about the summation of multiple waves and the possible loss of their identities, while in another, superposition is about the nonlocal nature of a single, unified wave.

As an analogy regarding the summation of waves, one can hear this phenomenon. Musical notes arrive at our ears at a particular frequency. Each resonates with a unique, audible tone. A note played alone is only a tone. But add two or more together and we create a chord. Each tone retains its identity in the sound but combines with the other to become another, richer sound. A fuller, more complex, and more beautiful sound, something above and

Summing to One - I / Superposition

beyond the individual tones that went into it. Interestingly, all music can be formed by the addition, subtraction, and transformation of only sine waves. If one has fooled around with a Digital Audio Workstation on their computer, one quickly realizes how all music can be created by superposing sine waves of varying wavelengths. By combining sine waves, one can create new waveforms including saw, square, and triangle waveforms. Each waveform possesses its own unique and unmistakable sound. The technical name for this is Fourier synthesis. Fourier analysis is about teasing apart waves into their fundamental tones, while synthesis is just the opposite.

Mathematically, the superposition principle states that the sum of two or more solutions to a linear equation is itself a solution. Translating that into a quantum's states, if it can be in *either* of two states, then it can be in *both* at the same time.

By toying with wavelengths of light, a Mach-Zehnder interferometer can beautifully illustrate superposition and the wavelike nature of *individual* quanta. The experiment also demonstrates their inherent randomness and indivisible nature.

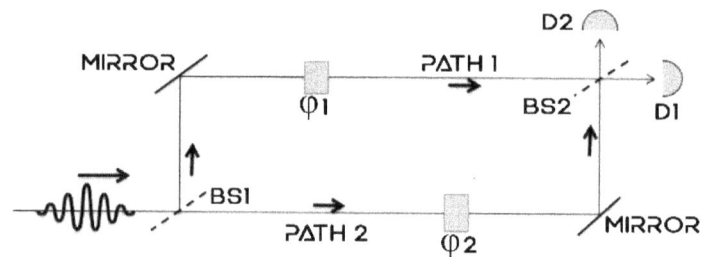

From the bottom-left, a laser beam of light shines it into the experiment where it meets a half-silvered mirror. This "beam-splitter" (BS1) cleaves the light ray in half, transmitting and reflecting it onto two different optical paths. Mirrors are set up on each route to redirect the light to cross itself where a second beam-splitter (BS2) may be placed. This second beam-splitter is necessary to recombine the transmitted and reflected light by transmitting and reflecting it *again*. Each path can then be equipped with a "phase shifter," ($\varphi 1$, $\varphi 2$) a device that acts as a spatial modifier, shortening or lengthening the paths taken by the light by a single wavelength or any fraction thereof.

The purpose of the phase-shifters is to allow us to manipulate how the relative wave phases will align at the second beam-splitter. Placed behind the two possible exit routes out of the last beam-splitter lies two photodetectors — D1 and D2.

With or without the second beam-splitter, a laser beam of monochromatic light and no change in path length, the experiment is set up such that half the light will arrive at D1 and the other at D2. This is because the first splitter cleaves the light in half. With the second beam-splitter in place, the phase shifters can then be set up to send all the light to either D1 or D2 or any fraction in between. The phase shifters accomplish this by altering the "length" of the path taken by the light, which in turn affects what part of the wave meets and interacts with the second beam-splitter. The word 'length' is here in scare quotes because the phase shifters don't alter the path length but instead do just as their namesake implies... shift the light's phase such that it is *as if* a change in path length has occurred.

To send all the light to D1, the path length of one branch is altered such that the second beam-splitter *recombines* the light moving toward it. That is, the light that the second beam-splitter both transmits and reflects does so such that the light headed toward D1 is *in phase*, while at the same time, the light it recombines heads toward D2 is out of phase. Again, by altering the "length" of either path, we can create an interference effect such that all the light will go to one or the other photodetector or any fraction in between.

Now, we go quantum and prepare identical photons and send them through the experiment one by one. Although they are individually unpredictable, we will find that they are statistically predictable. But first, to see how photons will behave all alone, we remove the second beam-splitter and ask: will a single photon be transmitted or reflected by the first beam-splitter? After many trials, what is found is what theory predicts. Half of the photons will be detected randomly at either photodetector with 50-50 probability. On any single trial, even with identically prepared photons and an unchanging set-up, there is no way to predict — no matter what — where the photon will end up. Quantum processes are intrinsically, and truly, *random*.

We also cannot tell "which path" the photon took. Was it on this path in |this⟩ state or on that path in |that⟩ state? Who knows? At this point, all we are certain of is that the statistics of their impact points agree with the formalism's prediction and are randomly distributed with 50% arriving at D1 and 50% at D2. A result in line with what we'd expect with full laser light and no second beam-splitter.

Revealing itself in point-like impacts, collapse as a reductive process, it almost looks as if a particle of light is randomly traveling the experiment. However, its particle-like collision into either photodetector need only cement the fact that quanta are unified bundles of field energy; units behaving as units do — whole and complete. We only ever detect an entire photon, never a fraction of a photon. They always collapse, all-or-nothing, into either/or detector—never both. We prepare and send one photon through and retrieve one at the end of the experiment.

Now, let's re-insert the second beam-splitter and ask ourselves a question; can we — by "altering the path lengths" as we did before — create the same results of having all the light (all of the photons in the quantum case) go to either/or photodetector? In other words, will a single photon — the so-called "particle of light" — incorporate information concerning both paths when it reconvenes at the second beam-splitter?

The answer is yes. Just as we saw with the full beam of light, the only way this result could be achieved is if the single, solitary photon follows *both* paths to *interfere with itself* at the second beam splitter. A result which illustrates, without any doubt whatsoever, that superposition is actual, physical, and real. And even though in the interim, we cannot know where the photon is by "seeing" it, we can nevertheless inferentially infer with certainty that it occupies and is on both paths simultaneously. For only by incorporating "which path" information gathered by traversing both paths simultaneously can the photon "decide" where it ought to go. The single unit of EM field energy, the photon, "splits" at the first beam-splitter, where it carries on to be reflected back by opposing mirrors toward itself. Here, at the second beam-splitter, it undergoes a final interaction where it interferes with itself. The result will always be in line with what the formalism predicts.

As an indivisible unit of field energy, the photon cannot split, and yet it divides at the first beam-splitter and goes both ways: just as our drop of water turned circularly expanding wave did. If we believed the photon was a particle, this would make no sense. But given that it's a wave through and through, *the photon is both transmitted and reflected by both beam splitters.* In transit, all the while, it retains a "nonlocal" coherency across itself and possesses a unified nature that "overleaps space."

Again, because it's a single unified object spread over a certain distance, the photon 'knows' where the photon is. As an individual, it is always-already instantly connected to itself, no matter how large it grows or distantly it disperses. Furthermore, the photon's particle-like collapse does not confer upon it a complimentary particulate nature but is instead only expressive of the odd character of collapse. An oddity explainable only by the quantum's ability to "jump" and the detection screen's constitution as a lattice of atoms.

Again, the single photon recombines at the second beam-splitter, where, from both paths, it is transmitted and reflected. By starting in the D1 configuration, all photons will arrive at D1. Then, by adjusting *either* path length, we can achieve with single photons what we did with "solid" laser light. This means that the photon remains a unified whole, even as its resonance splits and extends across the whole experiment. When recombined, it will interfere with itself and will go to whichever photodetector we send them to. Prior to interacting with a photodetector, the photon must incorporate data about both paths to "decide" into which detector it will collapse. Thus, while traversing the experiment, the photon is in a superposition of taking both paths simultaneously. And although we cannot tell with any certainty, we can infer that this situation remains true of the experiment *without* the second beam-splitter.

It's helpful to analyze the previous discussion in terms of *states*. Think of each path available to the photon as a particular state it can be in and remember the superposition principle: if a quantum can be in one of two or more states, it can be in all of them at once. So, the photon, before it's recombined is in a superposition state of being in both |this> and |that> state. This is what it means to be in a true superposition with respect to spatial extent or "position." But quanta can also be in a superposition state with respect to any one of their measurable properties, and this includes material quanta as well. For instance, electrons are known to spin, their wavefunctions are in a superposition where it has a component of spin in [this] direction and a component in [that] direction.

Note how a plane wave propagates out from the slit. As it refracts around the walls of the opening the wave begins to propagate spherically. As such, the wavefunction that would describe such a wave would have a vector component moving North-West, a component moving West, and yet another moving South-West and so on. The wave nature of quanta is what allows them to superpose, and any quantum property one can measure, can be in a superposition.

The signature of a superposition state is the interference phenomena that emerge after repeated trials of some proper measurable quantity. If a quantum can be in this or that state, the superposition principle has it that it can be in both at the same time. A composite quantum, however, is somewhat different. As an entangled whole, it can be in a superposition but the entangled quanta that make it up cannot and will have limited or "mixed" states, such that they are definite with respect to one another but are, prior to the act of measuring, unknown. The act of measurement will confer on us what they were. Entanglement, like a measurement, will limit the state of the composite quantum and define what an internal component can be. That is to say, the individual quanta making up a composite state will be in either |this> or |that> state, not a superposition of both. Therefore, *the individuals that build up composite quanta cannot themselves be in a superposition*, although the whole they form can. Noting this now will help us solve the measurement problem later.

As a reminder regarding this section's title, when two quanta *superpose*, they create a single, novel field distortion and thereby *sum to one*.

"I would not call entanglement *one* but rather *the* characteristic trait of quantum mechanics." - Erwin Schrödinger

CHAPTER XIII

Summing to One - II | Entanglement

That which makes the quantum extraordinary is its behaviour. As a wave, its solitary dynamics already goes against common intuition. Their ability to interact with themselves or quantum jump into new states is astounding. But what of systems harboring two or more quanta?

QFT says that when two quanta come together, they exert forces on one another, and through this interaction they *entangle;* after which they will behave as a *single* quantum. Until they are again disturbed, they will maintain an instantaneous link to one another, no matter the distance of their separation or the time transpired between interactions. The two share a single quantum state, and when the state collapses, say, by measuring the spin of one, it instantly affects the other. Even if the other doesn't know what you measured it assumes a correlation to the property measured.

Think of two separate quanta propagating toward one another in the psi field. As these electrons pass through and by one another they interact, however slightly, such that when they leave the interaction location they carry away information about the other. Their wave shapes become more complex after the interaction. As we saw with the pond of water, the configuration of the quantum field they share is in a new state. The quanta of said field are in an *entangled* superposition.

The states of the two have merged and are now instantaneously bound to one another such that a measurement on one will instantly affect and cause a real physical change in the other. Because the observable one wishes to measure is undetermined prior to its measurement, this effect cannot be used for superluminal communication and so does not violate relativity. Instead, it violates something called Bell's inequality. Named after its progenitor, John Bell, his inequality is a staistico-mathematical theorem that relates whether quantities (or observables) are correlated. His theorem is a statistical proof and not even necessarily quantum. It's said that if the inequality is "violated," it proves that the observables we endeavor to measure are related such that they share a nonlocal connection. As Bell's theorem predicts and experiments have shown, his inequality is indeed violated. As such, the effect of entanglement has been both mathematically and experimentally *proven.* In *Quantum Mechanics and Experience,* physicist turned philosopher David Albert writes: "What Bell has given us is a proof that there is, as a matter of fact, a genuine nonlocality in the actual workings of nature, *however* we attempt to describe it, period."[*]

How Bell arrived at his inequality was by trying to find a way to resolve the EPR paradox. The "paradox" being that quantum theory predicted that entangled quanta were nonlocally and instantaneously connected. Which to be fair, at the time seemed absurd. To Einstein, entanglement implied a sort of "spooky action." His disdain for it is what actually brought into the limelight. In a paper he and two junior colleagues wrote, they began a crusade to illustrate the incompleteness of quantum theory and came up with an ingenious thought experiment — now known as the EPR paradox for Einstein, Podolsky, and Rosen — meant to show that quantum theory would allow for some kind of nonlocal effect; an instantaneous communication of information, propagating signals faster than light and violating relativity. As it would turn out, it does allow for a nonlocal effect, but relativity is not violated. It's in the unitary nature of entangled quanta — their qubit — that solves the conundrum for there is no information, signal or otherwise, being transmitted across space. Entangled quanta share a single qubit. Measuring a property of one of the entangled quanta does not affect the other but by so doing we learn something about it because the are in direct correlation with one another. In measuring one, the binary nature of the qubit snaps and each correlate oppositely. It does not matter which property you measure; the binary opposing correlation will hold. Instantly. Entanglement does not involve action at a distance, spooky or otherwise. Today, thanks to Bell, we know better; nonlocality is real, entangled quanta behave as a quantum would.

Entanglement illustrates a common misconception about QFT; that it deals only with the incredibly small subatomic zoo. True, but only partially so, given that fields and individual quanta are spatially extended, QFT doesn't simply deal with the microconstituents of reality but its very fabric: the total tapestry. The supposed "quantum realm" is not another world but this very world. The task we have set ourselves — and will remedy in the section on decoherence — is explaining how the counterintuitive tapestry of the quantum can bring about macroscopic reality.

Entanglement leads to and is closely related to the concept of nonlocality. Nonlocality is sometimes taken to mean nonlocal affectation, such that one object can affect another, without touching, across an arbitrary distance.

---

[*] Albert, D. Z.: *Quantum Mechanics and Experience.* (1992)

However, nonlocality more deeply refers to the unity of the system under consideration. Quanta as waves are already "nonlocal" objects and entanglement is best understood as the unifying merger of two or more quanta into a novel, single system.

Thinking of the quanta that enter into a relation as retaining their identities leads to the notion that entangled quanta can have a nonlocal and instantaneous affect on one another. That effecting a change in one will *cause* the other to undergo a similar physical change across whatever distance. This view is wrong.

To acquiesce the correct one, we must be careful in how we define these interrelated processes. Entanglement is the process by which two or more quanta interact and become one although each keeps a semblance of its own identity. That is, if it was going one way before the interaction, it will continue to do so after.

Consider two quanta that meet in the middle. One arrives to the collision point from the bottom and the other from the left. Here, if they exert forces on one another, they will entangle such that each — although moving through the field as if nothing substantial happened — carries off a portion of the other. More concretely however, by entangling, the two now share a quantum state, one that can be represented as a single field configuration.

Nonlocal refers to the temporal and spatial extension of a unified quantum system. Nonlocality refers to the instantaneous correlation that develops between entangled quanta; the feature that — regardless of spatial separation — enables them to behave as a single system. If entangled quanta affected one another across a distance, information could be sent superluminally, breaking relativity. But entangled quanta don't affect each other, the are instead parts of a common global state. A global state which can be well defined even if the parts that form it are not.

Before we measure one of a quantum system's many "observables," they are indefinite or undetermined while also being restricted to their historical genesis.

Say you want to measure a quantum's spin. As we know, it is in a superposition of having part of its vector in |this> direction and part in |that> direction. That is, having not yet perturbed the system we know it to be in an indefinite state of realizing both. If the wavefunction maps onto reality than all possibilities are realized actualities. Now, the act of measurement — any interaction in fact — will reduce the state, collapsing it such that whatever one finds to be true of one, the other will correlated. The act of measuring one of the entangled quanta will cause a real physical change in the other quantum, an instantaneous action that is no action at all.

With regards an information ontology, entanglement may just be the process by which the universe generates information *ex nihilo*. Let me say that again, entanglement gifts us a model that allows us to coherently consider how the universe may have gone from zero to one; of how something came from nothing. This, of course has profound implications.

In his *Programming the Universe*, quantum computation scientist Seth Lloyd remarks that the operation of the quantum version of the controlled-NOT logic gate allows "quantum mechanics to create information *out of nothing*."[*] This is because the entanglement of two qubits *generates* a third qubit of information that was not present in the initial two. The entropy of the whole is raised, seemingly from nowhere, by the entanglement unification of many parts. Later, we will speculate about *creatio ex nihilo* — how the Big Bang may have come from *nothing*.

So far, we have only been discussing the subject matter in the form of an ideal case, that of a free field quantum theory — of single quanta fluctuating in their respective fields. But entanglement is all about *interaction* and no quantum — or any physical object for that matter— is so desolate, so isolated that it's completely alone, cut off from everything and devoid of any grounding relations. Even a single quantum evolving in a superposition will entangle with its vacuum field in its lowest energy state.[†] And as quanta interact and entangle, they grow and are further localized (decohered) by their environments. But we must keep in mind that composite quanta, when **maximally entangled**, achieve the same status that characterizes the solitary quantum. That is, individual quanta get together to form larger structures which in turn are so unified through the entanglement that the behave again as a quantum.

To be sure, there are degrees of entanglement where maximal means that the new system in question can be treated as a single quantum defined by a wavefunction. Likewise, there may exist any degree of ratio between entangled quanta such that one or the other may possess more affective "weight."

The implications of entanglement are far reaching. "One upshot from all this is the suggestion of universal entanglement," writes Art Hobson, is that "every quantum might be entangled with other quanta, and some of these entanglements might extend over macroscopic and cosmological distances. It is interesting to contemplate, then, the implication that appears to arise from nonlocality. Whenever a single quantum interacts here on Earth, some other quantum entangled with it and perhaps lying a great distance away, adjusts its

---

[*] Lloyd, S.: *Programming the Universe*. (2006)
[†] See Hobson for more.

quantum state. Arguably, every move you make causes an instantaneous microscopic quiver across the universe."

As an unnecessary aside, there exists another kind of instantaneous action in physics that can be expressed as this fact; a photon takes the same amount of time to cross the street as it does to cross the entire universe. But this is really just a trick of relativity, for when something moves at the speed of light it doesn't travel through time but only space. If photons could carry clocks, they would be frozen in time. This means that when you look up and swallow that star's light with the black hole that is your pupil, so far as the photon's concerned, it made the trip in zero time, that somehow, the the photon left its mother and landed in your eye at the same time. Even though it traveled billions of light years in distance, with regards it own frame of reference, it took zero time.

Now, bring this in close and codify it into terms of knowing consciousness. In your local environment, realize that by this mechanism, you and the objects of your perception are instantaneously interconnected across distance. The information a photon imbibes is timeless, allowing mind to know what-was as what-is. And this goes not only for the photons being absorbed by your eyes photoreceptors but is also true of the photons reflected from you and that go to others, the period of their vibration — frequency — are how light-waves carry with them a unit of time but do not themselves "experience" time.

We know through wavefunction collapse that the quantum demonstrates a unity that overleaps space. Now, through the entanglement experiments of Aspect as informed by Bell's theorem, we know that entangled quanta possess the same unity that overleaps space.

As a reminder regarding this section's title, when two quanta *entangle,* they create a novel field distortion and thereby *sum to one.*

The implication is clear: time and time again, quantum theory — while entertaining individuals — *unifies the whole of reality.* It grants to us an analogical method of seeing the mereological relation of how a part — a whole in and of itself — relates to a larger whole. This allows us to honor our individuality while also enabling us to see our common ground and origin as we are all equally expressions of one and the same Thing.

" Hobson, A.: *Tales of the Quantum.* Oxford University Press. (2017)

C

"After silence, that which comes nearest to expressing the inexpressible is music."
- Aldous Huxley

CHAPTER XIV

Chimes of Kaleidoscopic Color | The Music of the Spheres

Quantum chromodynamics is the theory of the strong nuclear force and the colorful fluids that "glue" together the cores of atoms. The maelstrom marble hearts of which — protons and neutrons — fall under the rubric of *nucleons*. A proton — a composite whole formed by a multiplicity of entangled quanta — possesses a single unit of positive electric charge and is described by a single wavefunction.

The proton constantly "shifts phase," producing a cloud of virtual photons more commonly understood as a unit of positive electric charge. With this, the proton captures and tethers a negatively charged electron to itself, cloaking itself underneath. Neutrons, possessing no charge, contribute only to the mass of the nucleus. Both protons and neutrons — as wholes — are fermions and intrinsically "spin," the axis of which can remain fixed or precess about a particular direction. That they "spin" means that these involuting tempests fold into themselves in a complicated way. A way we shall consider in detail later.

As for the malleable marble hearts of atoms; in a space less than a femtometer across and near the speed of light, an unfathomable amount of almost massless, blobs of '**quark**' field energy slide through and past one another and like jiving dancers — through a ceaseless exchange of virtual **gluons** — pull one another back together before they could ever part.

A threefold ratio of color quarks constantly turn into one another and "share" gluons as they do so. A dynamic dance that confines them into "flux tubes" of their own creation. This (almost) massless dance gives rise to true mass as something dense, something tangible. It creates the most "solid" part of the atom, the orb-like nucleons, which again can be modeled and described by singular wavefunctions and treated as a single quantum of a unified field. A proton field say.

Let's dive inside the protons and neutrons themselves to see how a spherical-like, pulsating orb of fluid energy keeps itself together and gives rise to mass. In diving deeper, we will meet a colorful cast of strange quanta and some novel field properties. As we go, and despite there being no real or representative basis for them, physicists have attempted to aid our failing phenomenological intuitions by giving to the exotic concepts of chromodynamics sensual counterparts. As is in the name, 'chromo' refers to color, the name given to the novel kind of charge that was uncovered.

The material quanta of chromodynamics are called 'quarks,' a name taken from James Joyce's irredeemably difficult book *Finnegan's Wake*. Each quark comes with a few qubits of information regarding its spin, energy, and a fraction of electric charge. Beyond this, each possesses a sliver of a new kind of charge, aptly called *color charge*. Color charge is very much *like* electric charge except its operation involves momentum, spin, and distance and is therefore much more complicated than the simple phase shifts that define electrodynamics.

Like electrodynamics, color charge is a quark's capacity to produce and respond to the phase shifts of other quarks, each one of which produces a virtual color gluon. In electrodynamics, depending on the direction of the phase shift, charge comes in two types conveniently labeled positive and negative. As we know, like signs repel and opposites attract. However, in chromodynamics, there are three 'types' of charge that interrelate with each other in a much more complicated fashion. Physicists have labeled these red, green, and blue, each of which is "carried" by a corresponding color quark.

When a red quark shifts phase, it "turns into" a blue quark *and* — conservation of energy required — color gluon. Unlike virtual photons, color gluons not only respond to quarks but also to one another thereby making the internal structure of a nucleon incomprehensibly complex. Gluons are responsible for keeping individual nucleons like protons "glued" together as well as gluing together distinct nucleons themselves. That is, gluons are the glue inside individual nucleons and the glue that glues them together.

In fact, in a maximally entangled atomic core, there remains little distinction between which part is a proton or neutron as the global whole is only ever a maelstrom of quark and gluon energy. And as we move up in energy, the fundamental object — the nucleus — only ever remains the same. That is, it remains a quantum of energy in the nucleon field that structures the hearts of atoms. Having greater energy only begets a greater effect upon the surrounding EM and gravitational fields such that it will consequently surround itself in the apt amount of electrical energy in the form of electrons.

Consider, we know the total core of any atom simply as "the nucleus." This nucleus is one and the same for every elemental atom and is indistinguishable from every one of its self-same others. Every nucleus of every oxygen atom contains eight protons and eight neutrons and could easily be exchanged, without ever noticing, with another. As a global whole, this nucleus has a kind of weight or effect upon the EM field such that it will eventually capture eight electrons and become a proper oxygen atom. However, the nucleus itself, when properly entangled is but one resonant distortion of a single field, call it the nucleus field.

Different quanta of energy in this field are representative of the various atomic weights of atoms but they are all energetic shapes of the same "nucleus" field. An analogy may be helpful. Imagine every planet as a sphere of pure iron. Some planets will weigh more than others, but they are all made of the same stuff. Those that weigh more have greater gravitational reach, that is, a stronger effect on their locally encompassing fields. A similar situation takes place in the heart of atoms. All are various levels of quark and gluon energy, as a whole, *made* of the same stuff. This is how entanglement sums the many to one.

To return to the quarks, there exist six *flavors*: up, down, strange, charm, top, and bottom. We will only concern ourselves with the up quark and the down quark, as the other four are inherently unstable and decay into the up and down quarks rather readily anyhow. Adding to this, the up and down quarks are all that are required to make up the nucleons of atoms.

We know not how many quarks are needed to form a solitary nucleon, only that there must be a threefold ratio of them to do so. Two up quarks for everyone down quark are required to form a proton. When the proton captures an electron and this trinity of fundamental quanta completes the most basic atom, hydrogen. Note, *only three material quanta are necessary to make up all heavy matter in the known universe*. All 'solidity' in the universe comes from the binding and kinetic energy of confined quarks.

Consider the heart of hydrogen; a single, solitary proton, an energetic dynamism as pulsating marble. Not only do quarks "have" color charge, but they also possess a degree of electric charge such that a proton will possess a unit of positive electric charge. We know, two up quarks and a single down quark conspire to make up a proton. An up quark has a positive electric charge, but only 2/3 of the value of an electron. While a down quark has -1/3. To properly dance with an electron, a proton requires a positive electric charge of 1 - unity. Thus, 2/3 for an up quark plus 2/3 for another up quark minus 1/3 from the down quark gives us a single integer of electric charge. In this way, our solitary proton can capture and dance with its electron and balance itself out.

Above and beyond the "nucleus" field, a proton plus electron forms hydrogen, the simplest chemical element, and their entanglement allows us to consider the atom as a whole as a single quantum of some larger-order field. That is to say, it is a specific amount of matter-field energy that is characterized as a wave.

Aside from color charge, three novel concepts characterize quantum chromodynamics: **color confinement** and **asymptotic freedom**, and **chiral symmetry breaking**.

Color confinement means that quarks cannot be isolated and observed alone. They are necessarily confined, only ever and always showing up in pairs. For instance, the "jets" we see in collider experiments are indicative of the point where it becomes more energetically favorable for a quark/anti-quark pair to manifest rather than for the original quark to carry on.

At the same time, quarks must obey a principle known as asymptotic freedom. One might wonder, is having to obey a freedom a contradiction in terms? Not quite. Asymptotic freedom tells us that as the quarks get farther apart from one another the force felt between them grows until they reach a critical point and can get go no farther from one another. At this point, the two quarks reverse direction and come barreling back into one another. It's as if they are confined to live inside the bounds of a rubber band, forever gyrating to and fro. Now, you might think calling a confining principle 'freedom' to be a bit of a misnomer. But what asymptotic freedom is referring to is the fact that as the quarks come closer and closer to one another the force begins to vanish such that when the wave packets of the quarks are passing through one another there is effectively no force at all. Quarks are "free" when they come together but once they fly past one another they begin to feel the color confinement begetting of their charge and begin to feel attracted to each other.

Finally, chiral symmetry is invariance under a parity transformation. Put simply, this means that independently rotating left-handed and right-handed components of quarks will make no difference to the theory. When this symmetry is spontaneously broken quarks acquire mass.

Electron Clouds & the Molecular Structure of Materials

To learn of the heavens, humanity had to develop an eye that could see beyond time and into the heart of matter. It was only by employing logic and rational mathematics that we began to tease truth from Nature. A monumental achievement, the Pythagorean theorem — by showing hidden numerical relationships in the geometry of objects — revealed a mirroring link between the mathematical domain and the physical world.

It was the first indisputable *proof,* the illustration of which caused an immense sociological change. Mathematical proof, as indisputably evidenced by Pythagoras, allowed once murky beliefs to desire truth. Of course, mathematical knowledge is not the only valuable form of knowledge, but it does form a type that is difficult, if not impossible, to dispute. Further, mathematics allows for the generation of knowledge if and only if it is coupled with the natural world.

Now, millennia later, we continue that tradition and manipulate numbers through logical rules to unerringly guide reason such that it allows us to envision a world too small to see. Today, true to the spirit of Pythagoras, we understand the mathematics of music and atomic physics are strikingly similar as they both describe the behavior of waves. Armed with the scalpel of pure mind, it would seem nothing is beyond the scope of our mathematically augmented eye.

We have seen that light is a wave in a field and claimed that all quanta are indeed waves, including material quanta. So how exactly do waves gather to become electrochemical atoms? How do electrons, the quanta that cloak atoms with their clouds, make molecules and create chemistry, all while providing the crucial step that leads to truly solid macroscopic objects?

Mathematically described, the atoms that constitute matter arise from the underlying laws as dynamic and beautiful multi-dimensional objects. By varying certain parameters and binding them by geometrical constraints one can solve Schrödinger's wave equation and uncover the standing wave patterns of electron orbitals as they emerge as beautiful and discrete units. The geometrical constraints are more commonly called boundary conditions and they dictate the harmonic frequencies the system can have. As we know, a guitar string has two boundary conditions while the atom has only one – the nucleus.

The Schrödinger equation deterministically describes the time-evolution of a matter-wave in a field. And when we bind the electron to a proton and solve the equation, the stunning shapes of the quantum world emerge. As Schrödinger stated in his Nobel lecture: "The atom, in reality, is ... the phenomenon of an electron-wave captured, as it were, by the nucleus of the atom."[*]

When one solves this equation by varying the input values for three differing quantum numbers, the kaleidoscopic shapes of the quantum world begin to appear. The principal quantum number designates the energy of the electron's shell and as this increases so too does its distance from the nucleus. The azimuthal quantum number determines the electron's angular momentum and dictates the shape of the orbital. The magnetic quantum number dictates the orientation of the orbital. Finally, a fourth quantum number exists; the spin quantum number was not yet known to Schrödinger.

---

[*] Schrödinger, E. R. J. A.: Nobel Lecture. (1933)

Its discovery was made some years later by Paul Dirac when he spent some time relativizing Schrödinger's equation to be consistent with special relativity.

Now, although they describe a continuous and smoothly varying field, the shapes are discretely different and, in some places, even seem to disappear. What are we to make of this image?

The electron is its own quantum and can only carry energy determined by its frequency. Adding specific amounts of energy to an electron does not multiply it, but instead excites it into a different state, giving it a higher frequency. They "jump" between orbital configurations by the energetic emission or absorption of photons if and only if those photons possess an energy that matches the discrete difference in the electron's potentially possible states. The electron is not truly bound to the proton. It carries the opposite charge of the proton and so "wants" to stay close, but it would almost equally be as happy as a free electron.

To understand the discrete, quantized orbital shapes of hydrogen's electron, we must consider the mechanics of standing waves and the resonant frequencies of vibrating strings. In one dimension, imagine a guitar string that is bound at both ends. This securing of the string sets up a periodic boundary condition and forms a geometrical constraint on the system. By plucking the string we set in motion a standing wave and the boundary condition sets an integer-valued limit on how many wavelengths one can fit on the string.

From this we can see that when a guitar string is plucked it vibrates smoothly and continuously, however, its geometrical confinement leads to it having discrete or unchanging properties. The wavelengths of the string are said to be quantized; each wavelength is continuously variable but discretely different from the next.

A string vibrating in the 2nd harmonic, with one full wavelength, will have a single node in its center where the displacement value will always be zero. In a standing wave, the string cannot vibrate in a different motion on either side of a node. The geometry of the situation requires discreteness and whole numbers because only certain frequencies can sustain themselves in the available space. This is what de Broglie referred to when he said: "Determination of the stable motion of electrons in the atom introduces integers... That suggested the idea to me that electrons themselves could not be represented as simple corpuscles either, but that a periodicity had also to be assigned to them too... The conditions of quantum stability thus emerge as analogous to resonance phenomena and the appearance of integers becomes as natural here as in the theory of vibrating cords and plates."

As de Broglie observed, one can view standing waves in two dimensions by vibrating and thereby setting up a resonant frequency on a plate covered in sand. Again, geometrical constraints and frequency dictate the patterns we see. Finally, adding in the third spatial dimension we begin to see the clandestine beauty of the music of the spheres.

Oscillating, kaleidoscopic movements of electrons dance, twist, and spin to cloak the nucleus in a haze of electrical energy, its single operant boundary condition. You see, with respect to atoms, the boundary condition is a single center point of attraction (the nucleus). This point also forms a node, a place where the field value is zero and as the electron grows in energetic complexity, more of these nodes come into being just as if we were to add more wavelengths to our string. In fact, in three dimensions, nodes are upgraded to *nodal surfaces*. The beautiful shapes that emerge are slivered partitions, snapshots of portions of the electrons themselves appearing to us as three-dimensional, spatially extended standing waves pulsating about the atom.

To be fair, normally we are told that visualizations like this show an enclosing region where a particle is likely to be found should we choose to interact with it. The Born rule is an interpretation that was quickly formed after Schrödinger had published his equation. Born, caught in classical times, wanted to interpret the wavefunction in particle terms and his rule seeks to treat the wavefunction as a mathematical abstraction whose values give us the probability density of finding the electron there.

We do better when we realize the shapes described by the mathematics really do correspond, at least partially, to the electron as a whole and are not just the expression of a surface. The volume occupied by the standing wave *is* the entire spread-out electron. With this, Born's probability density interpretation of the wavefunction is now to be understood as an energy-momentum-charge density distribution. It represents the strength of field energy where the electron is most likely to *interact,* rather than it as a particle will "be found."

The first viable and explanatory QFT — quantum electrodynamics (QED) — was birthed in mature form when Paul Dirac uncovered its landmark equation. By seeking a relativistic equation for the electron — an equation that encompassed SR and QM — he ushered in the field of QED. Dirac's equation — like Maxwell's before him — had much more to say than initially appeared

---

˙ Quoted in Brooks, R. A.: *Fields of Color.* (2016)

on the surface. By synthesizing a relativistic equation for the electron, Dirac revealed the electron's anti-quantum — the positron.

As is somewhat commonplace in theoretical physics, he mathematically proved the positron's existence — the electrons identical albeit oppositely charged twin — years before its experimental discovery. The positron is a wave in the same psi-field as the electron except it shifts phase in the opposite direction and therefore "has" its charge swapped—a kind of "hole" in the field. The mathematical power of QED is astonishing as it has been experimentally tested to the finest degree. The theory's prediction for the electron's "magnetic moment," agrees with experiments to an unrivaled degree and is thus our physical theory *par excellence.*

With Dirac's contribution, the theory had become almost fully formed, but it wasn't without its problems, however. Infinities kept showing up in some of the calculations, but they were dealt with in a way that made the theory somewhat more realistic. To deal with them, the mathematical program of **renormalization** was introduced. A process that restricted quantum field theory to an effective theory by reducing the energies considered to possibly actual ones.

"It is by logic that we prove, but by intuition that we discover. To know how to criticize is good, to know how to create is better." - Henri Poincaré

CHAPTER XV

Indeterminacy | Heisenberg v. Schrödinger

In the mid-1920s, both Heisenberg and Schrödinger completed the quantum mechanical program. Each with a different formalism that would later be shown to be equivalent ways of expressing the same underlying phenomena. Heisenberg created an overly complex mathematical monstrosity called matrix mechanics. While Schrödinger, taking a cue from De Broglie, created the much simpler wave mechanics. Schrödinger's formulation held a clearer ontological significance as quanta are more easily intuited as waves. Nevertheless, both are towering figures in the history of quantum theory.

The characteristic nature of waves leads to and illustrates a deeply hidden postulate of quantum theory that relates certain "conjugate variables," Heisenberg's famous **indeterminacy principle**. Heisenberg's principle fits quanta with an interrelated range of uncertainty between any two conjugate variables. In popular literature, the principle is most easily explained in terms of a single particle. We are told that the more you can know of a particle's position the less you can know of its momentum and vice versa. Position and momentum being the conjugate variables.

But, as we know, particles do not exist. The fundamental object of QFT is a quantum described by a wavefunction, the evolution of which is captured by quantum theory's central equation — Schrödinger's. So, before we analyze Heisenberg's insight, let's familiarize ourselves with Schrödinger's equation as it is to quantum theory what Newton's force law is to classical physics. Schrödinger's equation defines the deterministic dynamism of quanta and can tell us how a quantum system will evolve if left to its own devices. In a word, like a wave in water.

Schrödinger's equation uses something called a Hamiltonian, a quantity that compactly captures a system's various energies, whether potential or kinetic. On the opposite side, we see how that energy evolves the wavefunction over time. As it turns out, where there is greater energy the wavefunction will oscillate faster and have a greater amplitude digging into or out of the field (whichever way you find helpful to picture a multi-dimensional, standing wave's amplitude). Equally, where there is less energy, the wave will vibrate slower. Now, we can already see the wavefunction possesses a range of what we could call velocities, as such, it will also vary in position. Furthermore, we see that, because quanta are waves, these two conjugate variables — energy and time — are interrelated terms that dynamically affect one another as the wavefunction evolves.

Returning to Heisenberg, his equation solidifies and expresses this relationship and places a fundamentally necessary lower limit on them. This lower limit is a minimal amount that can easily be seen to be represented by Planck's initial view of the quantum itself. Heisenberg is not limited to time/energy but is applied to all the wavefunction's conjugate variables.

All quanta possess these inherent interrelationships. Because quanta are waves and do not have precise positions or velocities and Heisenberg's principle places a lower limit on them, we can infer that a quantum always-already has an intrinsic "size." Simply put, quanta cannot be localized to such an extent that they become point particles. It would violate Heisenberg's principle if it could.

Further to that point, and applying the same logic to time and energy, a quantum possesses an intrinsic temporal extension. This is difficult to intuit as we've seen that quanta are created instantly. So how can something always-already possess temporal extent yet be created in an instant? I can only offer my intuition and could be wrong here. Here's what I think allowing both Heisenberg's minimal lifespan and Planck's instantaneity means. The instantaneity is a moment we can pick out after the fact and is a real movement in the universe. A true and total change in state.

Combining these insights together we see that the creation and annihilation of quanta is a smoothly evolving, probabilistic process, where somewhere along the line there exists an isolatable moment when it goes from |this> to |that> state. That moment we call "collapse." Although collapse purports to show a real change consider again what must be taking place. Simply, a morphing of energy exchanging and activating spatially inseparable fields.

To see Heisenberg's immortal contribution at work, consider again a wave in a medium and remember that velocity is a vector quantity such that its direction cannot be ignored. After passing through a slit, a wave spreads out circularly, the part unaffected by the edges maintains at velocity while the parts that diffract around them are slowed down. Now the wave is moving in multiple directions, each with interrelated velocities, at once — a true superposition.

Now, Heisenberg tells us that the product of a quantum's uncertainty in position and velocity is greater than some lower limit. The two terms "affect" one another in such a way that if one is high the other is counterfactually and equivocally low such that the product of the two remains above some certain value.

On the face of it, Heisenberg's principle has nothing to do with humans and our knowledge or ignorance of quantum states such as a particle's position versus velocity. Indeterminacy relations are a core feature of quantum theory and are therefore intrinsic to the nature of reality of quantized waves.

On a final note, Heisenberg's indeterminacy principle was anticipated by and resolves some of the ingenious paradoxes of the ancient thinker Zeno of Elea. His paradoxes of motion — that relate space and time — were motivated by a desire to differentiate between orders of time: the indivisible flow of real-time as grasped by direct intuition and the infinitely divisible and abstractly partitionable time as grasped by the mathematically armed intellect. These two temporal "orders" thus defined seem to be at odds with one another and Zeno's tale of a frozen arrow aims to bring their incompatibility into sharp relief.

Zeno aims to show that there can be no "instant" in time with respect to the evolution of a system. If you pick out an instant, you lose motion and there would be no information contained in that instant that would allow you to predict how the system would evolve in the next one. To achieve this, imagine an arrow suspended in mid-air such that it occupies "a place exactly equivalent to its length"˙ and freeze the frame. Zeno points out that if one knows the exact position of the arrow at an exact moment in time, one cannot discern anything of its trajectory. Is it moving? Is it falling? Was it let loose by a bow? Tossed into the air by a child? There is no way to tell. In allowing time to be cut up into frozen instants — t1, t2, t3 — motion is lost and in any discernable moment, the arrow is effectively standing still. If that's the case, then the arrow never moves at all.

Again, in knowing its *exact* position, we can discern *nothing* of its trajectory (or in modern terminology, its momentum). Although Zeno's thought experiment is the absolute case, Heisenberg's position/momentum uncertainty relation *proves* that this situation is impossible. Quantum mechanically, we simply cannot freeze the frame to discern an *exact* position. The quantum as a minimal unit of substance forbids this. Further, it shows that position and momentum co-define one another and that there exists a fundamental lower limit to this relation.

˙ Aristotle.: *Physics*. (Before Jesus/old-as-fuck)

Indeterminacy / Heisenberg v. Schrödinger

Does this imply that temporal instants do not exist and are meaningless? Not quite, it simply means that motion cannot be accurately defined in a moment and that a lower limit of temporal extension is required to define anything. That is to say, a minimum of previous "moments" must be taken into account to predict anything. And that is just what Heisenberg is saying, a minimum amount of spatiotemporal extension is required to relate position and momentum.

Now that we are familiar with much of the quantum's uncommon characteristics, let us turn now to the two most intricate quantum field theories. First, we will visit quantum chromodynamics and dive into the marble hearts of atoms. Later, we will visit quantum electrodynamics and probe the scission between matter and light and analyze the electromagnetic bonds that beget molecular structure.

> "A strange loop is a paradoxical level-crossing feedback loop."
> – Douglas Hofstadter

CHAPTER XVI

The Strangeness of Spin | Ouroboros & The Seed of Self-Reference

First identified as an internal degree of freedom belonging to the electron, we have seen that some quanta (fermions) can "spin." Spin appears as an independent physical parameter in the formalism of quantum electrodynamics and is revealed by Dirac's landmark equation that forms the heart of that QFT.

This we know of spin: that it is a strictly quantum mechanical feature with no classical analog. It shows up as an extra degree of freedom and inferentially observable property belonging to fermions. Its most basic feature can be summed up by saying that you have to go around twice to arrive at where you started. Spin relies upon and is revealed and necessitated by the symmetry constraints that structure Minkowski spacetime. That it requires spacetime symmetries for its ontic actuality, spin must be intrinsic to the very fabric of spacetime itself.

So, before we look at the wave picture of spin, let us first examine its origin as arising from the spinor field of Minkowski spacetime; a spatiotemporal scaffolding that possesses symmetry properties that demand a certain kind of primordial self-reference. Fermions are also "spinors" as they form the most basic individual units of this type of complex field. Spinors "spin" such that they fold back upon themselves and never really "go" anywhere.

To see how spin arises from the structural symmetry principles of SR remember that it relates space and time in such a way that they mix, and certain operations done upon this fabric must be accounted for and accommodated *by* the fabric itself. This, we know, turns space and time into a 4-dimensional manifold wherein either space can be "time-like" when required or time can be "space-like" when necessary. This is due to geometrical constraints that impose certain restrictions on said fabric. As such, symmetries relating to spatial translations, rotations, and boosts between inertial frames form these symmetries and it is because of these symmetries that spin emerges as a novel property.

The Strangeness of Spin / Ouroboros & The Seed of Self-Reference

Consider any mathematical point of a scalar field. Because it possesses only a magnitude, any rotation done around this point will remain symmetric and there will be no way to tell if a rotation has even transpired. The spin value of such a field is zero or "spin-zero."

Next, consider a point of a vector field such as the EM field. Possessing both a magnitude and direction, it "picks out" a preferred direction in space, one which must be accounted for when considering symmetric transformations. A rotation around a point in a vector field requires one full revolution to "turn back into itself" and for this reason, they are said to possess "integer spin" or are "spin one."

Finally, consider a spinor field, a point in which is a complex function. Just as with the imaginary unit i — the square root of negative one — was a difficult concept to uncover and coherently grasp, the nature of spinor fields also remains mysterious. Indeed, each point in a spinor field can be abstractly thought of as the "square root" of geometry such that a rotation about any point in a spinor field will require *two* full revolutions to return to itself. This strange capacity has revealed itself through the symmetries that structure Minkowski spacetime. That is to say, Minkowski spacetime is a spinor field and thereby possesses the properties that demand of its denizens that they possess a certain kind of primordial self-reference. These objects — spinors — fold back upon themselves in a wholly unique way such that they only ever turn into themselves.

$$♂$$

But just what exactly *is* spin and how do quanta as waves exhibit it? As quantum theory would have it, spin proves to be an elusive concept that is difficult to pin down physically.

Mathematically, we know that quanta possess a definite value of spin and that this value is quantized — a binary, either/or, this or that quantity. As with all quanta, an electron's spin orientation can be in a superposition where part of its wavefunction has a component of spin in $|this>$ direction and another component in $|that>$ direction simultaneously. As we know, when "measured" its state will decohere and limit itself to either one of these.

Although it is revealed to us by experiments and through the mathematics, we are unsure of how to "picture" it. Nevertheless, we shall try.

Classical intuition regarding spin tempts one to view the quantum as a tiny frictionless dynamo — a particle rotating about some axis — but this would lead to a violation of relativity as the "point-particle" would have to spin superluminally. If we understand the electron is a wave-packet of spread-out

The Strangeness of Spin | Ouroboros & The Seed of Self-Reference

energy-momentum-charge density, the violation is avoided.* Visualizing the electron as a torus-like, involuting, rolling wave-packet may stretch the imagination but the following picture that will be presented remains faithful to experiment and the mathematics.†

That is to say, spin is better understood as a way in which a wave rolls *while rolling over itself again.*

Now, how to get there? To begin to examine the wave picture of spin, let us imagine adding an extra degree of freedom to a waving photon; essentially, a circular aspect that, like a clock, can only rotate in either one of two directions. As regards light, this is called polarization.

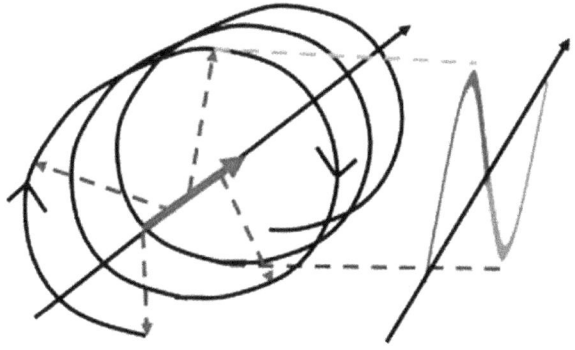

ADDING A ROTATIONAL ASPECT TO A WAVE'S PROPAGATION

Adding this aspect to an electron causes it to manifest a magnetic field with a strength proportional to it. Due to the discrete nature of spin, as clockwise or counter-clockwise, electrons sent through a Stern-Gerlach apparatus will deflect in one of two ways — either this way or that — with none ever landing in the middle. In the initial setup of the apparatus, electrons were deflected either up or down, so today we often refer to spin quantities as spin-up or spin-down.

---

* Sebens, C. T.: *How Electrons Spin.* (2018).
† Schmitz, W.: *Particles, Fields and Forces.* (2019)

The Strangeness of Spin | Ouroboros & The Seed of Self-Reference

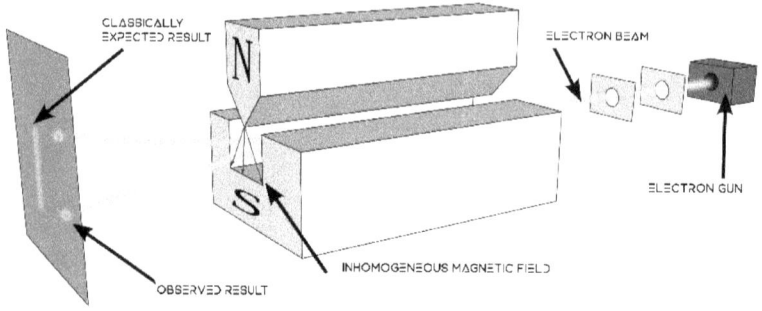

Mythological Meandering

Before we continue with quantum spin, I should like to make a philosophical detour and note what is meant by this section's title *the seed of self-reference*, pursue its ancient origins as a concept, and consider it as a necessity for certain forms of mindedness. As will be made apparent, the capacity for self-reference leads to paradox and seeming contradiction.

The Self is the space in which something might appear and is the structure that reveals itself when we become aware of our awareness. Not only do we perceive the world as it is given to us — this we share with all other knowing creatures — but we also perceive our perceiving and in so doing achieve the meta-cognitive state of self-consciousness. By coupling this achievement to our ability to willfully guide our thinking — while implicitly understanding that the window of conscious experience is temporally extended — we become self-determining beings. Through this curious structure of human-grade consciousness, we can isolate and construct an *identity*. An identity that is nothing more than a holographic projection of who we believe ourselves to be — a phantom phenomenon built up by possibly false beliefs as well as certain facts. This "ego" is not who we really are but is instead a mask that hides our true nature from ourselves and conceals the Divine essence that each of us, most deeply, are. To come to *know* ourselves as divine requires a transcendental mode of knowing.

Foregoing that for the moment and remaining in the time-bound mundane mode of consciousness, by employing the temporal extension of conscious experience — by retaining in the present a memory of the just-past *as just-past but in the present* — we see just what the self, as ego, *is*: a mental projection or epistemic object. This pattern integrity, this ego is then thrown into the about-to-happen. It is by this temporal function that in this endlessly rolling moment we call the present, the self as subject of experience comes to split, objectify, and know itself as both an *experiencing* and *narrative* self. The first is total perception, experience itself, while the other is a wholly psychological construct held up by the mind's eye and its capacity to remember. The narrative self is a rational concept — the image we have of ourselves — and one that strange loopily feeds back into itself to guide how we live our lives. Despite its essence as a sociological construct, it is only by recognizing its actuality that we can make choices that honor ourselves as a unique being. It is only with our storied narrative self in mind that we decide on the actions we make in life.

But again, it is only within the sliver of time that the experiencing self comes to see itself as a self and this — at least so far as conscious awareness is concerned — forms the seed of its self-reference.

As the master logician, Kurt Gödel so succinctly (and M. C. Escher artistically) pointed out, self-reference — the capacity for a thing to refer to itself — has an innocuous ability that leads to much contradiction and paradox. Self-reference also allows for — in fact *generates* — the involution of an **entangled hierarchy**; an abstract object known more commonly as a strange loop.

Two of those terms need a little clarification. In mathematics, an "involution" is a function or transformation that when applied to itself gives its own identity, a transformation that leads to its own inverse. In philosophy, an involution refers to the process by which an object "turns into" and upon itself. While a "strange loop" is a kind of abstract structure or feedback loop that doesn't actually seem to be a loop. At first, as one moves along the loop it seems to be "going somewhere" but after a certain period one ends up back where one started.

Colloquially defined by cognitive scientist Douglas Hofstadter, a strange loop is "an abstract loop in which, in the series of stages that constitute the cycling-around, there is a shift from one level of abstraction (or structure) to another, which feels like an upwards movement in a hierarchy, and yet somehow the successive "upward" shifts turn out to give rise to a closed cycle."˙

As we know, with regards the nature of the self as a person, Hofstadter states: "The self comes into being at the moment it has the power to reflect itself." For Hofstadter, the essence of the self is none other than a strange loop. With regards to human consciousness, this is a fundamental feature of mindedness and is without base.

Now, what is the historical genesis of this strange loop and seed of self-reference?

˙ Hofstadter, D.: *I am a Strange Loop.* (2007)

The Strangeness of Spin / Ouroboros & The Seed of Self-Reference

It is to be found in the most comprehensive image that represents Life itself — and is indicative of what I call the **Morality of the Mystic** — it is the snake that consumes itself. Originating first in ancient Egyptian iconography; Ouroboros is a mythologem that appears in many cultures across the globe. An image of a circle entwined; a snake devours its own tail. It is the first adumbration in the collective human consciousness of a strange loop; an object that folds back upon itself.

Bound in time, a finite symbol alludes to an infinite nature, Ouroboros represents the eternal efflorescence of the timeless now as it is. By swallowing its tail, it speaks of death and rebirth. But more than this, by turning in time, Ouroboros signifies incessant Becoming, but by going nowhere, is also expressive of pure Being and it is this endless cycle of turning in place that articulates the eternal return of Life itself. By turning in place, the serpent symbolizes the evolution of the universe and the fact that it only builds upon itself, within itself, as itself. Death and rebirth are pale understandings of this image, and its essence is best captured by the poet Jorge Luis Borges: "Time is the substance from which I am made. Time is a river which carries me along, but I am the river; it's a tiger that devours me, but I am the tiger; it's a fire that consumes me, but I am the fire."

But of its nature as consuming itself was best put by Schopenhauer: ""

Beholden to a more physical interpretation, one might consider the snake to represent the conservation of energy. More specifically, the "total energy" of the universe as it is only ever energy that builds within and upon energy, efflorescent evolution at its finest.

Finally, the involutive nature of the self leads us to understand the Morality of the Mystic. As we know, the standard waking state of relative consciousness is polarized by a subject/object dichotomy. To learn to survive in a hostile environment, the subject/object dichotomy that permeates relative consciousness is one of the first brain programs we install as children as we must learn to differentiate between what is self and not-self and recognize and differentiate between the self-as-subject and the self-as-object-for-others. All our lives this mental operating program will "run in the background" and inform perception. A functional operation that takes place in the subconscious before presenting the world to consciousness.

Above and beyond relative consciousness lies its transcendental counterpart. When the distinction between what is self and not-self is dissolved, this modality of mind comes to be known as a blissfully liberating epiphany. In a loopily strange moment where "consciousness turns upon itself towards its source," consciousness becomes its own object, and the dichotomy

---

˙ Merrell-Wolff, F.: *Transformations in Consciousness*. (1995)

is dissolved into one of undifferentiated Unity. By this, we learn that we are the entire universe itself but expressing itself at this moment, at this place and time, as a person — indeed, as you.

This is of course but another way of describing the liberating knowledge of the enlightenment experience and is part of what the image of Ouroboros is meant to show. That the thing known is identical to the knower. The thing consumed is identical to the consumer. Tat Tvam Asi — the gurus of the Upanishads say — Thou Art That. This is the morality of the mystic and the deepest understanding of the image.

Now, let's investigate paradox proper as other Ouroborisian knots (strange loops) exist. The most obvious of which is the language-generated strange loop known as *Epimenides' paradox*, or *liar's paradox*. Epimenides, a Cretan, was a philosopher-poet whose original poem *Cretica* read: "Cretans, all liars." Given that Epimenides himself was a Cretan, his statement implies a paradox. For if it speaks true, it is a lie, and if it is a lie, it cannot, by definition, be true. This turn of phrase is more easily seen by the negative statement "this sentence is false," and is the typical example given when considering the strange loop engendered by language. Put simply, if the sentence is true then it's false, and if it's false then it's true. By moving through the sentence, we seem to make progress, stepping up into a novel syntactical understanding but only ever find ourselves back where we started. As we have just seen, concerning the philosophy of mind, self-consciousness, and its consequential ability to self-refer, are paramount.

We all refer to our "selves" with the most common pronoun in existence, the double-entendre that is I (eye). As a conscious agent, who is it that "I" to which we refer when we say I? The answer is the seat of the self, the "that" which "is never seen but is the seer; it is never heard but is the hearer; it is never thought of but is the thinker; it is never known but is the knower." Hofstadter believes that it is this enigmatic capacity of self-reference that forms the archetypal prerequisite for mindedness. But this is only partially the case as this strange loop of the mind is only realized by sufficiently complex creatures. "Lower-order" creatures may lack self-consciousness, but that does not mean that its possibility is not available to them, or certainly, that they do not lack consciousness as such.

Hofstadter goes so far as to claim that he himself is a strange loop. That the "I" that refers to itself as itself is a strange loop. We can bring this insight into sharper relief by considering again the quarks of QCD. They are fermions that possess half-integer spin such that they fold into themselves *only* after

---

Brihad-Aranyaka Upanishad, III.VIII. Translated by Swami Nikhilananda. (1990)

making two full rotations (why two turns will be seen in the next section). This, as it turns out, is the defining feature of a fermion. In QFT, fermions couple to other fermions by exchanging intermediate interaction bosons. In QCD, the resulting, maximally entangled superposition state of confined quarks forms a proton. A proton which alone, when maximally entangled, *behaves as but another fermion*! This means that all singular, unified material objects sum to one to become *an individual fermion*. In our case, a human organism with an involutive property inherent to itself as itself. Again, every isolatable, material entity — every "solid body in space" — when maximally entangled within and as itself, is a fermion. Mathematically definable as a wavefunction; a single, solitary quantum state.

It bears repeating, every solid, self-sustaining material object, can be modeled as, and effectively is, a single fermionic quantum system glued together by the binding energy of bosons.

Quantum mechanical spin is a basic, fundamental feature of the universe. We see it as a strange loop that lies at the very bottom of the world, structuring spacetime itself, and is therefore a property inherent to all things — whether or not they express or exhibit it.

Now, not only do we observe self-reference taking place in the mind but also in everyday objects and structures. In artworks, like the famous image of M.C. Escher where two hands draw one another. Or in a piece of music like the fugues of Sebastian Bach that "end" by returning to their beginning. Or, most vividly, in the domain of logic as seen by Epimenides' paradox.

These differing art forms all capture a common property, that is, they are all strange loops. Concerning Escher, the strange loops of his art are beautiful illusions; drawings that visually bring out the very essence of a strange loop. For Bach, his fugues are real auditory phenomena that seem to change only to ever return to themselves. Similarly, a "Shepard Tone" is a sound built from superposed sine waves that create the illusion of an endlessly rising or falling tone that in reality never "goes" anywhere.

Finally, Gödel's **incompleteness theorems** — logico-informational strange loops — are the most devastating result to ever transpire to mathematical logic as they prove, without any doubt whatsoever, that no axiomatic system that *refers to itself* can ever be complete. Essentially saying that there always must exist something above and beyond the system itself, something that cannot itself be contained within the system but contains the system itself.

For Gödel, the human mind is the That which is able to see what is immanently the case. This is because it can simply see a contradiction in terms, without having to prove it. Knowing about the properties of the objects that enter into the relation of functional mathematics allows the mind to bear witness to truth without recourse to anything else. We see immediately and without reference to anything else that $2 + 2 = 5$ is false, and that it is simply not necessary to apply the rules of the axiomatic system that define these objects to prove it. We simply see it, immediately. It is for this reason, that consciousness is above and beyond the material world from which it is naively thought to be borne, that the great physicist Roger Penrose believes it to *not* be computational in nature.

An aside concerning the metaphysics of the day: it is because we are witnessing beings with the ability to refer to ourselves that we cannot be living in a simulation. The image of the world as a simulation is helpful to explain hidden facts about our world — such that it is always and ever One and that everywhere throughout it, it has the self-same substantiality and structure. But the usefulness of simulation as an analogical tool ends there for the very simple reason that simulations are not conscious; nor do they refer to themselves.

Even though the time in which we currently exist is referred to as 'the age of information,' the universe is — and more importantly, *is not* — a quantum computer. Certainly, our reality can be *modeled* on one as it certainly *does* compute information, and so is, in effect, a quantum computer, but Nature is not limited by this, and it doesn't force it to be one. The true nature of the quantum foam is such that it not only computes information and energetically interacts with itself through quantized fields, but it also harbors spirit and consciousness — two "things" that are not accounted for in any computational metaphysics.

Despite this, many wish to model the "mind" on a neurocomputational view, with "the brain" serving as an organism's central processing unit or CPU. The brain is seen as the "hardware" that underlies and begets certain psychological functions; its "software" programs. In the human bio-computer model, the primary senses are seen as channels of streaming information that define a type of relation as to what-is and what-is-not the self.

Although it may be a helpful analogy to see the brain/mind as a kind of rational engine, this image is both incomplete and quite simply false. Corroborated by Penrose, the human mind is above and beyond a computer.

Often ignored, or simply going unnoticed, the final nail put into the coffin of this neurocomputational model of mind was hammered down many years ago by the immortal Alan Turing when he showed that *self-reference in computation leads directly to incomputability.*

As regards mindedness then, the moment of self-consciousness' genesis marks the moment where we must enlarge our concept of what the universe is. Going beyond Gödel, that self-reference exists means Nature must be more complicated than a mere, albeit wholly fascinating, quantum computer. That consciousness exists to illuminate and know all this illustrates that the universe cannot simply be a computer, no matter how complicated, nor a simulation, no matter how vivid. Put simply, Nature outreaches us and will always be found to be *more than* our best models.

A waterfall with no origin, ceaselessly streaming; spin marks an eternal efflorescence, a forever fountain of constant involution and endless becoming. The only "thing" that can harbor spin as a phenomenon is the quantum foam as vacuum. But the primordial object that first embodies it is the living cell. Said differently, the signature, self-referring, strange loop nature of spin can be seen in the quintessential form of life as we know it, in the biological cell.

A living cell is an autopoietic, self-forming, self-interacting phenomenon. The most basic unit of life, its dynamics instantiate the simultaneous co-emergence of an inside and outside; defining primarily what is and is not the self. This relation can be seen as an endless *folding back upon itself to generate itself* making the solitary yet unified prokaryotic cell the first instance of a macroscopically visible and dynamic expression of spin.

As a unified object describable by a single wavefunction, the composite cell involves in its constitution many physical fields and probably reaches into and activates many as of yet unseen others. A primary thesis of this work is that the living cell forms the quantum of the mindfield. It is the minimal ripple of unitary energy that can disturb and thereby activate the timeless field of — always-already present — consciousness.

Said somewhat differently, regarding spin from a biological perspective, unlike a eukaryote, a prokaryote is without a central nucleus but nevertheless constructs itself for itself and is the most basic expression of spin because it exists by instantiating that structure. Just as a dressed electron is an oscillation in multiple fields that realize different portions of the quantum foam as carrier, the living, biological cell forms the quantum of the mindfield; its "minimal unit" that utilizes the multiplicity of fields of which it is constructed to break into a novel and far more exotic feature of the foam as carrier — the sentient sea and field of consciousness itself.

With respect to human mindedness, and most assuredly mindedness in general, the comprehensive structure revealed by the self-referential character of spin also reveals a truism with regards to consciousness as such; that it is an "emergent property" that "supervenes" on the processes that allow it to emerge in the first place. Here, supervenience means only that consciousness demonstrates downward causation and cannot be reduced to the processes, forms, and structures that are in place to bring it about. Consciousness is a "something" somehow above and beyond the materials that constitute it.

Macroscopic objects dictate the movements of their microconstituents even though those microconstituents form the basis for their very being. This mereological "part to whole" relation is far greater explained by the quantum theoretic picture of reality as every organism is seen as an entangled hierarchy, a strange loop not just in Hofstadter's sense, but in the quantum mechanical sense. In the quantum state sense, this means an organism possesses an *instantaneous* connection to itself across the whole of itself. To be sure, the propagation of signals in the nervous system does take time to reach the cortex for processing but this matters not *with respect to how conscious awareness is experienced in real time.* I will say it again, "the foot feels the foot when it feels the ground." The entangled systems of quanta present to and resonating in the mindfield field connect instantly over a distance. This, to my mind, brings to light a novel connection between a system's constituent parts and the whole that they create.

As only a strange loop is capable of, the moment of its "folding back upon itself" without ever actually "going" anywhere can also be considered as a way in which it "moves" in two directions at once. As we saw earlier, consciousness is inherently, always-already, bidirectional, "moving" two ways at once. The world as we find it impresses itself upon us as if our witnessing were a kind of gravity. At the same time, the well of our awareness reaches out from us to arrive at the objects of our contemplation. The environmental sphere so defined forms — an organism's *umwelt* — a realm of possible action where "consciousness" is the slivered periphery of this movement actualizing itself at this irretractable locus where it processes transactional information.

Sensually, we embody and *are* an incessant stream of information that flows into the body and couples itself to an inner drive that — like a ghost in a shell— can move it about. Each of us intimately knows our willpower by knowing what it is not and even if most of our cognition is processed subconsciously, what we know as will is that part of ourselves with which we are in total control. We may not will what happens to us, but we can will how we respond to whatever may transpire.

The whole point is that spin is a phenomenon that is notoriously difficult to pin down. Not just in the quantum sense, but also in the many ways in which it pops up in cognitive science and logic. It may not even be the case that spin and self-reference equate as I have been endeavoring to show. But certainly, they both possess many similar characteristics and I do see a benefit in understanding spin, and spinors generally — an intrinsic property revealed by the symmetric structure of spacetime itself — as the primordial instance of self-reference. And if the seed of self-reference is to be found at the bottom of

---

* Umwelt is the German word for "surroundings" or "environment" and is defined by Thomas Sebeok as the "biological foundations that lie at the very epicenter of the study of both communication and signification in the human [and non-human] animal."

the universe — endogenous to the baseless fabric of this vision — then it is not without reason that we find it so easily expressed in and by us.

My apologies for the overly technical exegetical detour and philosophical rumination on spin metaphysics. Let us return to the quantum and see how it may be that a wave can spin.

"Picturing" Spin by Twisting Rotation Again

Most quanta rotate but only fermions spin. Photons may be circularly polarized but electrons are "strange loopily" polarized. That is to say, both photons and electrons may possess a helical aspect or characteristic, but only electrons — fermions more generally — add to this helical degree of freedom a further degree of rotational complexity. It is this degree of complexity that allows "material" quanta (e.g., electrons, quarks, protons) to utilize their spin dynamics to become the building bricks that solidify structure to form the furniture of reality.

Essentially, it is their spin value that causes them to obey the Pauli exclusion principle. This in turn allows them to "stack up" upon themselves by having every unique electron occupy a distinctly different orbital shape. They sum-total situation to which every quantum contributes itself to results in larger and larger objects such as heavier atoms, molecules, and so forth.

Spin is regarded as a type of intrinsic angular momentum, a "quantity" above and beyond the momentum and normalized orbital angular momentum of non-relativistic quanta. What this extra degree of freedom allows for is a far more complex and dynamic waveform. Harboring it, a quantum wave can rotate around a particular axis and rotate around that rotation again. This extra element — that twists rotation a second time — is quantum spin.

Furthermore, the spin-statistics of quanta are what separate matter and force (interaction) quanta at a fundamental level and structure the fields they're in. The two types, as we know, are fermions and bosons and each possesses a unique spin property because of the field they resonate within.

Fermions have half-integer spin, cannot sum, and make up massive quanta; their wavefunctions are anti-symmetric. Bosons have integer spin, can sum, and make up the force-carrying (interaction) quanta; their wavefunctions are symmetric. If anti-symmetric wavefunctions interfere they cancel, while symmetric one's sum. We shall see why shortly but for now know that after a period an anti-symmetric wavefunction has not yet returned to itself.

One may remember from high-school chemistry that the first "shell" of an atom can only harbor two electrons. The two cannot be the same as they would cancel each other out. Let us see why.

The Strangeness of Spin / Ouroboros & The Seed of Self-Reference

The Pauli exclusion principle holds that no two material quanta can be in the same state, for if they were, their respective wavefunctions would interfere and cancel one another out. The reason is that instead of summing to a greater field amplitude — as is the case with the symmetric summation of boson wavefunctions — fermionic wavefunctions are anti-symmetric and will deconstructively interfere with one another resulting in a null field value for that location. A place where 1 + 1 will = 0. Luckily, this does not, and cannot happen.

As we saw in the music of the spheres, two electrons can occupy the first shell because they differ by a single quantum number, which designates their spin value. Having opposite spin, one being spin |up> and the other spin |down>, the "direction" that each electron will dig into the electric field will be opposed. With the first shell filled it becomes impossible to fit more electrical energy into it. This forces all other electrons that wish to aid in the formation of an atom to occupy states further and further away from the nucleus. In this way, by not allowing material quanta to have identical states, matter begins to grow, taking on size and shape.

Now, how to visualize spin? How might we envisage a seemingly impossible rolling involution of a propagating wave? Before we try, we would do well to remember that a quantum is a wave-packet of energy oscillating in multiple fields with a constant phase and possessing both real and virtual "parts." The "real" is its robust quality as it is generated by being its home-fields true resonance, while its "virtual" portion is but a mere disturbance of said field and borrows energy from the vacuum to "feel" out what else may be present to the field. Said differently, the "real" aspect of quanta originates by having its native fields resonate authentically. While conversely, the "virtual" component merely disrupts these fields, drawing energy from the vacuum to explore the broader context of them.

Ultimately, this entity is an energy-momentum-charge density — a blob of "energy substance" — that delicately displaces and warps its host field(s). It is this structure we should visualize as folding and twisting back upon itself and thus "spinning." Due to its inherent ghost-like unity, this higher-order structure remains a single quantum, a ripple in a universal matter field.

To begin, consider a Mobius strip. A strange doubly rotated shape that to completely traverse takes two revolutions. Aha! That sounds a lot like spin. If you've never seen a Mobius strip, one can be made in an instant. Simply cut a piece of paper into a long rectangular shape, twist it, and connect the ends. If you stick to one surface and trace your way around this object, it will take two full revolutions to return to the place you started.

The Strangeness of Spin / Ouroboros & The Seed of Self-Reference

Although fermions do not spin exactly like this, the anti-symmetric, helical aspect of fermions has them behave in this way for if they returned to themselves after only a single rotation, they would cancel themselves out. But right where the electron would meet with and cancel itself out, its twists rotation again such that it may continue unimpeded. Bosons travel only around a circle, with no added twist, and with symmetric wavefunctions, they may sum. This is the main difference in their spin characteristics. The circularly polarized (i.e., helical) aspect of light, of quanta in general, is called **helicity**.

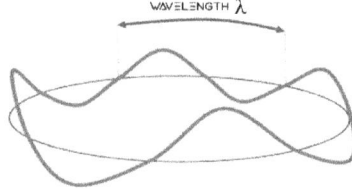

To see how a wave can spin about an axis and come back into itself without canceling itself out, consider one traveling around a circle. How would that look and how many complete wavelengths could we fit on it?

The answer depends on where the wave's phase would meet up with itself after completing a revolution. As the next image shows, a single wavelength will travel around a circle and continue to do so indefinitely.

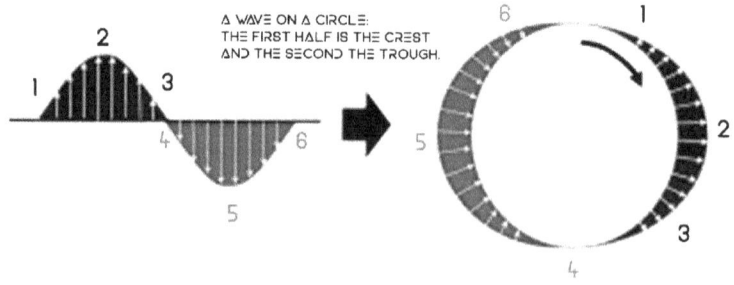

The Strangeness of Spin | Ouroboros & The Seed of Self-Reference

The crest-half of the phase occupies the first 180 degrees and the trough-half the latter 180 degrees. In fact, any integer-valued wavelength will behave like this. It is a symmetric wavefunction as after one revolution — one period or oscillation — it returns to where it began. It is this that makes a boson a boson and is why they are said to possess integer-valued spin.

However, concerning bosons, this is called circular polarization as they don't utilize the geometry of the Mobius strip and so don't twist. If they did, they wouldn't be able to sum, and if they can't sum, then lasers can't exist.

Now, what about a fraction-valued wavelength? What about half-integer spin? What if we were to add multiple half-wavelengths such that we fit two crest-halves and a single trough-half onto the circle? The result is that after one revolution it does not meet up with where it started. Instead, a crest will meet another crest and their phases will cancel each other out and the wave will perish.

But what if we allowed this wave to go around again? What if we gave the circle a twist, turning it into a Mobius strip? Now, after one revolution the crest will continue into its trough and after completing two full revolutions — the 1½ valued wave — will finally return to where it began. That the twist can be accomplished in one of two ways — clockwise or counter-clockwise — defines a quantum's chirality as either "left" or "right" handed.

Belonging only to fermions, it is this capacity that leads to the Pauli exclusion principle. As we know, Pauli's principle excludes the possibility of having two electrons in the same state. As we saw in the Music of the Spheres, four quantum numbers are required to describe a single atom-captured electron.

In the ground state of hydrogen, only two electrons are said to be able to occupy the orbital. If these two electrons were to have symmetric wavefunctions, all their quantum numbers the same, they couldn't exist. "Spinning" symmetrically would cause them to interfere destructively with one another and they would cancel out. So, at the very least, one must have an opposite spin value. Then, and only then, can two electrons occupy the same orbital near the center of the atom as each will be "spinning" through the other, activating the opposite "side" of the field, rather than interfering and canceling out.

The fascinating implication of all this is this: anti-symmetric wavefunctions — spinors, all of which are fermions — have a field amplitude on *only one side of the Mobius strip*. Such is why two can occupy a single one. In fact, they are each other's opposite.

Finally, the quantum mechanical property of spin distinguishes three types of fields: two for bosons and one for fermions. Possessing no circular polarization, the scalar Higgs field is a boson field with a spin value of zero while the EM is a vector field with circular polarization and a spin value of one. Fermions are spinor fields with half-integer values of spin.

Every electron everywhere is a definite individual, these reclusively solitary quanta are all indistinguishable from one another save for their inherent spin. But they can come together to form larger objects by way of entanglement. As a nucleon captures more and more electrons it grows bigger as each new electron must oscillate in its own space. Only by this quantum phenomenon can electrons remain individuals but come together and behave as a single unit. Even in billions of-atoms-large solid objects.

As objects grow, by taking advantage of Pauli's principle, matter as we know it begins to take shape. All atomic and molecular structure is defined by this single principle. And all electrons know where all others in their local region are by way of their virtual photon clouds. This electromagnetic glue binds atoms to atoms and molecules to molecules where eventually, somehow, a structure of such complexity may form such that it can actualize an inactive, previously unutilized — but always-already there — field; the mindfield.

Dictating Design | Non-Abelian Phase Invariance

Change Without Change; a mystical Parmenidean notion if there ever was one. Toned by a temporal intuition, it attempts to express the sacred knowledge of the unity of reality; that All is One and that this movement is nothing but endless flux: "wholeness in flowing movement." Despite the seemingly illusory nature of temporal evolution, what is fundamentally real has never actually changed; it is the movement of movement itself.

Bohm, D.: *Wholeness and the Implicate Order.* (1980) One of my favorite books.

Only the timeless now exists. Never has a moment existed apart from it. Step into and be washed away by this torrent of the present and realize the true nature of the world as a ceaseless unceasing. Change Without Change attempts to state, attempts to capture the fact that since its inception, the totality that is the 'universe' has only ever expressed itself and evolved as itself. That the endless multiplicity of forms and the seemingly separate objects we know are but different manifestations of one and the same underlying ground. A "ground" that has forever and unalterably remained the same.

The aphorism "change without change" attempts to express the fact that despite all the indelibly obvious change the universe has indeed undergone, it has at the same time, somehow, remained unchanged, "underneath" it all.

Like mass without mass, change without change is a phrase borrowed from Frank Wilczek and it signifies an exquisite way to express a seemingly inexpressible contradiction: that this universe can change without actually changing. To be fair, Wilczek was speaking of physical symmetry. But I am here using it in a philosophical sense with regard to the very nature of the universe.

As for physics, Philip Anderson may have said it best when he remarked that "it is only slightly overstating the case to say that physics is the study of symmetry." This is because, through the canonization of symmetry, it may be — as we saw with spin — that *all the laws of physics can be generated by symmetry principles.*

Now, exactly what is symmetry? Put simply, a symmetry represents an invariant property of some object that undergoes a transformation or operation of some sort. Invariant means only "does not change." Rotating a sphere by any amount leaves the object changed but unchanged. Thus, the concept of symmetry involves identity and difference and relates the two. Like a strange loop, a symmetry is a transformation that brings an object back into itself. If no transformation — save the object's own identity can bring it back into itself — it is said to be asymmetric.

The larger a group of transformations an object is able to undergo while remaining itself, the more symmetric it is said to be. For instance, there are infinitely many ways to transform a sphere such that it remains a sphere. It can be rotated by any amount, reflected, or translated, and will always remain a sphere. A cube, however, must be rotated and flipped in ninety-degree increments to return to itself and therefore the set of transformations under which it remains symmetric is far less than that of the sphere.

A set of symmetry transformations forms an algebraic structure called a group. A group often singles out an overarching invariant; the thing that does not change throughout the change. The Galilean group (acting on classical "absolute" space and time) of transformations isolates temporal intervals and spatial distance as invariants whereas the Lorentz group isolates the proper time interval of relativity.

Many symmetry groups contain subgroups, and a symmetry is said to 'break' when it is reduced to one of them. The immediate consequence is that new features and invariants emerge. For instance, the electroweak symmetry unifies the weak nuclear force and electromagnetism. When it is broken at low energies, we can see that both electric charge and weak hypercharge emerge as distinct invariants.

Famously, every continuous symmetry found in nature is not only expressive of a conserved quantity via Noether's theorem but that this quantity is the thing that necessitates the actuality of interaction phase (gauge) fields. Mass-energy is conserved through time translation symmetry; linear momentum through spatial translation symmetry; angular momentum through rotational symmetry; electric charge through the following "gauge" symmetry about to be discussed.

Now, given this preamble on symmetry, we may be prepared to wrap our heads around the most notorious concept in quantum field theory: non-abelian gauge symmetry. But if we are to honor what it actually represents, we would do well to call it a phase, rather than gauge, symmetry as that is what it is actually about. As the famous Chen Ning Yang of Yang-Mills theory repeatedly exclaimed: ""

Let's attack this difficult concept by decomposing it into its composite words. "Non-abelian" means that something does not commute and that the order of operations matters. Expressed as a simple equation − $a + b \neq b + a$. That is, rolling a die forward and then to the left does not result in the same situation as if the die is first rolled to the left and then forward. In this example, the two situations are said to not commute.

As for "gauge," the word is a leftover fossil of history. Initially introduced by Herman Weyl when he attempted to unify electromagnetism and GR. Weyl sought to change the "gauge" or scale of quantities so that they would remain invariant under such spacetime dependant operations. Einstein quickly showed him it led to errors. But the idea stuck. Kind of. Now we know "gauge" invariance actually pertains to the phase of a quantum's wavefunction. So, the proper name would be phase invariance for it's the phase of the wavefunction that remains invariant after operations. And again, it matters in which way you order the operations.

Symmetry, we know, involves identity, difference, and their relation. In other words, an object, the transformations, or operations it can undergo, and the way it relates to itself after said transformations.

The Strangeness of Spin / Ouroboros & The Seed of Self-Reference

Now, the requirement of local phase invariance leads to the conservation of electric charge. That is, imagine the wavefunction is a circle, indeed, mathematically, π is used to describe the phase of the wavefunction. Perform a rotation on the circle. Say, 90 degrees. After the rotation, the world remains unchanged. Change without change. When we do that charge is conserved. Now, when we make the symmetry depend on spacetime coordinates, where the phase shift takes place in space and time, that is we force the symmetry to be local, as they say, as it must be because the objects are indeed locatable events, this distinction without a difference must be propagated away. It is this requirement that couples the psi field to the electromagnetic. An electron's phase shift from 0 to 90 scratches the EM field ad sets up a corresponding wave in that field.

> "And in those days people will seek death and will not find it. They will long to die, but death will flee from them. - Revelation 9:6"

CHAPTER XVII

How to Kill Zombie Cats

After all the seemingly endless preamble we are now able to examine and hopefully resolve the most troubling remaining question in quantum theory and find a meaningful way out of the measurement problem while holding to a realistic interpretation of the theory.

To do so, we must analyze the quintessential quantum experiment — the proverbial double-slit. First used by Thomas Young to prove the wave nature of light; it is now used to illustrate many of quantum theory's most counterintuitive principles. Young used pure light to perform his demonstration, whereas the quantum analog uses singular quanta — this can be a photon or electron, or even sufficiently entangled composite quanta like protons, neutrons, atoms, or molecules.

The experiment is set up in the same way as Young did so many years ago. Two slits are placed in a barrier and light is shone through to land upon a screen some distance away. Because the light leaving the slits will interfere with itself, it will show up in bands. Now, the quantum version of the experiment remains just as simple but uses single, solitary quanta instead of pure light. Given this, we will only ever acquire point-like impact data points from the experiment, and we will therefore need to perform the same experiment many times to reveal the wavelike quality that they retain.

The experiment also serves as an empirical investigation into the evolution and resolution of the equation which forms the basis of quantum theory — Schrödinger's equation. The dynamics of this equation are put on vivid display by quantum theory's most well-known anecdote, the tale of Schrödinger's zombie cat.

In his famous example, Schrödinger has us imagine a steel chamber, in it is placed a tiny bit of radioactive substance, a Geiger counter, a vial of poison, and a cat. The Geiger counter is there to register radioactive decay. If it registers decay, it will trigger a device that smashes the vial of a gaseous poison and kills the cat. Schrödinger reminds us that we must secure the poison and Geiger counter from the potential tomfoolery of the cat. After a given amount of time, the probability that our atomic nucleus will randomly radioactively

How to Kill Zombie Cats

decay is 50/50 and the state will have evolved into a superposition actualizing both simultaneously — the cat is a zombie, both alive and dead.

However, opening the box to discover which state the cat is in causes the state to collapse. Although theory implies the cat was just actually existing as a real zombie, we only ever see an alive or dead cat. At first glance, the tale *seems* to predict an irreconcilable situation as it clearly elucidates an irretractable absurdity. The cat cannot be both alive and dead. But as we shall see, this is not actually the case. The true state of affairs is far more subtle and even simple.

To arrive there, let's first analogically consider the double-slit experiment alongside Schrödinger's horror story. Both the tale and experiment have three parts: radioactive decay as a quantum or quantum process, a Geiger counter that serves as a which-path detector (this will be explained in what follows), and a cat that exemplifies the final, global state of the system which in the experiment is revealed as the ending mark on the detection screen.

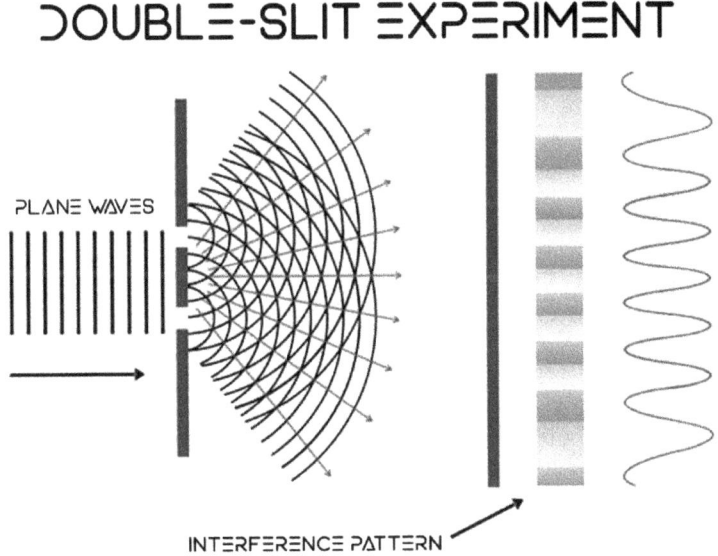

For our purposes, we shall use the photon as a stand-in for our radioactive quantum process. To do so, we simply reduce the intensity of the light such that only single photons are sent through, one at a time. The quantum is directed toward a doubly slitted wall where a detection screen sits some distance beyond it. At the barrier, a which-slit detector may be placed to see which slit the quantum goes through. This "which-path" is akin to the Geiger counter as it functions to register an either-or possibility, which, in the case of

How to Kill Zombie Cats

the cat, signifies whether radioactive decay has or has not taken place. Finally, the function of the detection screen is to absorb the photon and amplify its point of impact — whether it lands behind either slit is indicative of the cat's well-being. On the left, the cat is alive, on the right, not so much.

Now, if quanta are particles, they will *only-ever* go through either/or slit — never both — and after accumulated trials will produce a single band of impact points behind each slit. The cat will always appear as alive or dead. But if quanta are waves, they will go through both slits and produce the interference pattern of many light and dark bands; the same pattern Young observed when he demonstrated the wave nature of light.

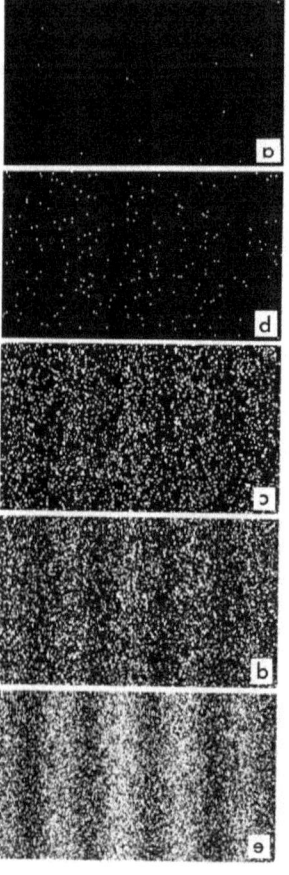

It bears repeating, that no matter what, all quanta will arrive at the detection screen "like a particle." Their quantized nature forces them to deposit their energy all or nothing. It is only after accumulated trials that a pixelated interference pattern will begin to emerge that is telling of their wavelike constitution. Quanta always appear to arrive "as particles" because of their unitary nature and interaction dynamics, not because they *are* particles. That an interference pattern emerges at all tells us one thing — that every single photon *goes through both slits, interferes with itself, and deposits its energy at a location where it has interfered constructively.*

It is worth noting that it is precisely due to these issues the experiment is thought to show the wave/particle duality inherent to the nature of reality. Taking both as actually existing forces one to interpret the theory through the lens of the Copenhagen interpretation. An interpretation that treats the world non-realistically as it holds that reality is undetermined — or worse, non-existent depending on your brand of Copenhagen — *before* observation. As the quantum traverses the experiment, it is a "wave of possibility," and after reaching the other side and interacting with the detection screen, "becomes" a solid particle. This ontology is unacceptable as waves and particles form mutually exclusive concepts.

Seen under this light, the immortal physicist John Wheeler referred to the life of a double-slit-experiment-crossing quantum as a "great smoky dragon," wherein we know its origin (tail) and its endpoint (bite) with precision, but its body remains veiled by smoke. Accordingly, Copenhagen intuitions saturated by the completeness of Newtonian mechanics and its point particles, hold that we do not know what a quantum is or what it is doing while traversing the experiment. And so, it is believed that while it travels the experiment, one cannot consider it *to be* anything. In other words, Copenhagen tells us the quantum has no reality whatsoever during the time it takes to move across the experiment. Again, clearly an unacceptable view. As we've seen, a proper, ontologically considerate, and realistic view is to understand the quantum as an actual wave of energy spread out in the field that — because of its unitarity, and the atomic structure of the detection screen — must collapse whole when interacting with the screen.

Now it is here that collapse becomes a contentious issue. That is if we accept the admittedly *ad hoc* postulate of quantum state "collapse" into our theory. This postulate is *ad hoc* because it is not found anywhere in the formal theory but is, time and time again, revealed by experiment.

Without collapse, the smooth evolution of Schrödinger's equation predicts a deterministic outcome where the new state is a new superposition of states. Unclothing this calculation down to its bare bones and holding to its vivid implication literally implies that the world — which is a universal wavefunction — with every interaction, "splits" into indefinitely many, almost identical, but never-to-interact, *worlds*. Each world forms an evolving branch of one and the same wavefunction resulting in a world where this result obtains, another with that result, and so on *ad indefinitum*.

Many-Worlds is arrived at by taking literally the fact that the Schrödinger equation always evolves into superpositions. It is baked into the theory, and this is where the interpretation gets its alluring power and value. But honestly, it's a bit much. Accepting that there may exist a copy of you orthogonal to this domain that is still dating their high-school sweetheart or another one of you who is a famous actor or whatever is demeaning and robs this Reality of its value. Many-Worlds means many mo' problems.

As we know, this strange conundrum we have been considering is *the problem of definite outcomes* and is indicative of where the maths and experiment do not seem to match up. If we want to hold onto this, our shared reality, and recognize that we only ever observe a single, definite outcome, never a superposition of outcomes, we must admit that, although collapse is not found in the formal mathematics but is what we *actually observe* in real life, it *must* be a physically tangible feature of the world.

A caveat — just as Many-Worlds is *implied* by the dynamics of Schrödinger evolution, wavefunction collapse is *implied* by Planck's quantum hypothesis but this does not mean that it's correct. Problems with accepting collapse remain, like why all the mass-energy suddenly shows up at one, point-like place rather than remaining spread evenly about the wavefunction itself. However, I find the more elegant view is to accept collapse as a real feature of reality as it keeps to its unity while allowing us to isolate the interaction point of relation. That is, collapse allows us to see how objects, quanta especially, are individuals even though the ontology of quantum fields so revealed becomes one of the ever-fractaling layers of kaleidoscopic complexity.

Seen most vividly as a function of magnitude, many layers of reality simultaneously exist to build each other up. The result of which is the emergence of novel objects that are, at the same time, larger-order entities that invert their influence from the "top-down." We are unimaginably more complex than — but nevertheless owe our existence to — atomic and molecular structure. And so, in a way, *we are* these coexisting domains that gather to form one and the same world. It is my personal belief that through meditation and other forms of entheogenic-assisted perception — psychedelic phenomenology brought about by the ingestion of a mind-manifesting molecular substance — we are capable of being made aware of these domains as the domains themselves and witness their majesty first-hand. After all, these layers of reality are indeed *in us* and serve as **complexity platform**s that build our organism's bodily form, the 'thing' that precipitates consciousness.

Apologies for the tangent, but it's worth noting that, in either view, Many-Worlds or Objective Collapse (the two I see as most closely resembling reality), the quantum never was, is, or will become, a particle.

To return to the problem at hand, inadequate resolutions to the measurement problem continue to spawn endless "interpretations." And regardless of which interpretation one wishes to advocate for, the fact that interpretations continue to proliferate is unacceptable. Either something is deeply wrong with the theory or somewhere along the way, we made a mistake. Possibly a misreading of some feature or a collective misinterpretation that continues to go unnoticed, or possibly, maybe a resolution has been offered but is still ignored because many hold a specific interpretation dear and do not wish to alter it.

As I will argue, hiding in plain view, such a resolution has been found, and it is as simple as can be. In a word, the original analysis of his own equation was flawed. Put simply, from the outset, Schrödinger misled himself, and forever after we have been pondering and arguing over the existence of zombie cats. As is often the case, the truth is simpler. As it turns out, the measurement state of Schrödinger's cat does not actually predict a zombified superposition state represented by a dead/alive cat. Instead, it predicts that the outcomes of experiments will be correlated to match observation. It predicts *either* a dead cat *or* an alive cat and that this observation will match what the Geiger counter reads. Yes, it's that simple.

This resolution has come to be called *the local-state solution to the problem of definite outcomes*. Its insights are simple but profound. To see it, one must only accept the unity of the quantum that leads to its ability to collapse and also understand the experimental evidence for why we know it happens and why it doesn't violate special relativity. And furthermore, how it (or something like it) allows for entangled quanta to behave as a single quantum — that is, composite quanta collapse across the spatial extension of their constituting subsystems in an instant — but again, this does not allow for information transfer or to be used as a signal.

Before "measurement," a quantum's properties haven't been defined by having a distinction made. It is not that the quantum is ill-defined. Quanta really is flowing movements of efflorescent potential that establish their existence at the periphery of Being and Becoming.

As for the local-state solution, the simplified version is that the measurement state — the global state of cat plus Geiger — is not itself a superposition state but rather *a superposition of correlations* that establish themselves when the states of its subsystems *entangle*. That may sound trivial but it's not. And Schrödinger himself may be forgiven as entanglement wasn't proven to be a real feature of the universe, nor even well understood, until well after he had dreamed up the thought experiment.

The standard analysis of Schrödinger's cat is flawed. It does not predict a zombie cat but instead predicts — with 100% accuracy — what the *correlations will be* with regard to the *entangled* subsystems that gather to comprise it. In other words, the measurement state is an *entangled superposition* state. This global, measurement state involves a "which-path" detector *and the* final detector (state of the cat), and it is how they come to unify themselves that defines them. Essentially, entanglement is the key to resolving the paradox for it sets up actual, causal, but as-of-yet undetermined relations. That is, when the state of the Geiger counter entangles with the state of the cat, it causes them to assume *definite* but not-until-the-box-is-opened *undetermined* states.

What the measurement state *is not* is a state in which the Geiger counter registers both decayed *and* not decayed and the cat is, at the same time, both dead *and* alive. One or other of these situations actually obtains and will be seen as perfectly correlated with one another but have not yet opened the box, we do not and cannot know which. A quantum **mixture** is a state wherein a system possesses *actual but as of yet indeterminate properties*.

Now, to truly understand why this is the case, and why we ought to accept the local-state solution, we must analyze completely the double slit experiment.

Schrödinger's Affair — The Double-Slit

Much has been said of the double-slit, but we have not yet performed it quantum mechanically. So, for the moment, let's move backward and begin by setting up a resonance — a single photon — in the EM field and send it towards a doubly-slitted barrier.

As we know, to create a photon, an electron must drop an energy level as it pulsates about an atom, and its standing wave shape must re-configure itself by "shifting its wavelength." Caught in the boundary-defining pull of the nucleus, the electron as standing wave can only "jump" between discrete configurations. This "jump" will register itself in an adjacent field as a precise and discrete amount of energy — a photon. Basically, a waveform slides between fields to ring the other out.

The orbital changes an electron undergoes will "give" the photon all its properties; its "most general" directedness, energy-momentum, polarization, and wavelength, which, if in the visible spectrum, will be seen by us as a color. Finally, we know the photon's speed to be c as the elasticity of this most basic field is immutable.

Let's go on another tangent! As an exciting aside, we should also note that a quantum is *instantaneously everywhere* once created. Wait; what!? But yes, an incredible assertion first pointed out by de Broglie, in an instant, the energy of the entire universal EM field gains a quantum and is thus instantaneously everywhere; as large the universe, however large it may be.

Because of its wave-like structure, although highly unlikely, any quantum in question will have a non-zero probability of being anywhere and everywhere in the universe. This is because the value of its wavefront can never be certain. Somewhat paradoxically, the photon is also highly localized in the instant of its creation. Being created in the space where the electron dropped an energy level, it moves out from here at the speed of light, taking with it all the respective information about the "surface" it left/was created by.

Consider a macroscopic correlation to intuit better what is going on here. Imagine a single drop of water falling into the ocean. This drop is our quantum, a discrete, specific amount of substance—be it water or energy. The very moment it hits the water, we can say that the entire ocean's overall volume is *instantaneously* one drop more than whatever it was previously. Again, *instantly*, over the whole thing.

The quantum version is a bit different in that it only deals with energy and not volume, but one can see in the analogy the sense in which a single drop loses itself to become the whole instantaneously. Secondly, returning to the ocean's surface, we can observe how the drop affects the body of the water as it moves into it. As it hits the surface, it begins setting up a ripple — a wave — while losing its identity as it merges with the water. But wait, although our drop of water has combined with the ocean and lost itself, it has become a circularly expanding wave whose continuously varying wavefront as a moving displacement of water cannot be well-defined and essentially stretches out unto infinity.

Another insight to glean from this analogy is that it matters not what creates the wave but that it is the water's elastic properties that will dictate the speed at which the waves within it will travel, as well as define the relation between frequency and wavelength. And furthermore, no amount of additional energy can speed them up — no matter what. Whether it be a drop of water, a rock, or a bowling ball, all waves created by these will move at the same rate, albeit with different amplitudes. For the very same reason, the EM field serves as a kind of universal speed limit. The elasticity of the EM field is a fundamental, unalterable property of the field itself. It is what it is and that is why nothing can go faster than light. One can put more and more energy into the EM field and create more energetic photons, but the waves will never speed up, they just can't.

Back to the double slit!

We set up our apparatus, and one at a time, send photons towards the slitted wall. Immediately after leaving its source, the wavefunction of the photon will begin to expand circularly while retaining a most "general directedness" (that is, to the right) as it moves towards the slits. If it were a material quantum, such as an electron or proton, as an expanding wave, we would see Heisenberg's uncertainty relations in action as the propagating waveform evolves to have a multivalued directionality, velocity, and momentum. I say a material quantum as photons are unable to have multivalued velocities.

Now, we cannot "see" any part of this dragon because it is veiled by smoke. We cannot see a photon reflect and bounce off a barrier or diffract around a corner. But we can and will infer what must be taking place at the slits for our results to make any sense at all.

Finally, the photon reaches the slits in a completely coherent state. As a precursor to the section *defining decoherence*, a quantum is "coherent" because it has evolved into its current superposition status unperturbed and thereby remains a unified resonance of EM field energy. It is a spatially extended, non-localized waveform that "occupies" multiple places at once and *has not interacted with anything else*. A quantum only retains its **coherence** as a unified object, if and only if, it has not yet interacted with anything else.

When it does, if it were to interact with another quantum say, they would entangle, and both would lose their coherent status. By such an interaction both states, aptly, decohere. Therefore, any action or interaction that causes a quantum to lose its coherent superposition status is called decoherence.

As we shall see, decoherence — the loss of superposition coherence — is a facet of the entanglement process that allows us to make the quantum-to-classical transitory leap. Adding to that, understanding it is of great importance when we attempt to solve the problems associated with Schrödinger's zombie cat.

Again, as it begins to press up against the slitted wall, the photon-as-wave finds its way through both. The photon is in a coherent superposition of coming through both. The unified nature of the quantum wave allows it to do this. In more detail, upon reaching the slits, some parts of the global waveform will reflect off the wall, but other parts of the wave will "feel" their way through, taking the path of least resistance. When leaving the slits and exiting through the other side, the single photon will diffract around the edges to become *two* circularly expanding wavefronts — one from each slit.

It bears reinforcing, this increasingly complicated mess and constantly evolving shape remain a harmonious quantum, a single increment of EM field energy, a "somewhat-distorted-by-the-environment" wave. And although it is unlikely, to be fair, it may be that the part of the photon that was reflected by the wall eventually collapses back into its source. However, we are only interested in the ones that make it through and exit the slits, as only these will mark themselves on the detection screen.

Now, leaving the slits, the photon's "now-two" wavefronts will begin to interfere with each other to create the signature superposition pattern. Yes, a *single* photon is superposing and interfering with itself! It remains in a coherent state while its novel phase relationships create an even more intricate pattern. It does not decohere because it has not entangled with anything else. And although the superposition pattern may look messy, it is as "coherent" and unified as it was before it went through the slits.

By its phase, the wave encodes — for itself, as itself — all the information about itself, but most importantly, *where it most substantially is* as it begins to press up against the final detector wall. As it begins to do so, the probability that it will interact with the detector, and thereby collapse, increases. Where the quantum has de-constructively interfered with itself, there will be almost no probability of interacting with the wall at said location. Whereas, where it has constructively interfered, it will have the greatest likelihood of interaction and will collapse at that location.

When the wave finally does collapse, it will *look* like a particle, but again, there is no particle to be found here, just the mark of an interactive exchange of energy. That the globally extended wave collapses to a well-defined point is due only to the size of the awaiting atom, not because it somehow becomes a particle. The energy that was our quantum now entangles to form the detector itself as well as begin the cascade of processes that amplified will show the location of collapse.

Again, wavefunction collapse "looks like a particle" because of the molecular structure of the detection screen itself as billions upon billions of atoms make it up. And its "surface," at this level, is to be seen as an effervescent sea of electrochemical energy, a bubbling pool of gyrating, semi-entangled electrons that are so intertwined that they are undifferentiable. Like how the water of the ocean is formed of countless H2O molecules, the surface of this electrochemical sea is smoothly varying but made up of innumerable electrons. Some are so close to their neighbors that they have entangled, while others are not so much.

Finally, when the photon's wavefunction begins to press into this sea at the places of its *constructive* interference, there will be a moment where one ensemble of nuclei-captured electrons will be ready to exactly accept the photon's energy and cause it to (wholly randomly I might add) collapse. Prior to this moment, there is no way, even in principle, to tell where the photon will "land." All we are certain of is that it will, and when it does, this energy that excites an electron will cause a cascade of electric movement, the signal will become amplified, and we will see the dragon's bite and know where the photon collapsed.

At this point, it may be helpful to switch analogies and use one of Art Hobson's illuminating examples. In his explanation of collapse, he equates the extended quantum with a rubber balloon filled with air and the detection screen with a bed of tightly spaced nails. Imagine too that this balloon is made of thick rubber so that it will take some force to pop it. The balloon represents the wavefunction as it incorporates the superposition pattern, and each nail represents a single atom in the lattice-like structure of the screen. As the balloon begins to encounter the bed of nails, there is a high probability that the initial contact point will pop it. But it's not guaranteed. As the balloon continues to press itself up against the bed of nails, encountering more and more nails as it does so, the probability that it will pop increases. But now, not just at the initial place. Any nail it presses up against could cause it to pop (collapse).

A situation akin to this happens with every collapse of the wavefunction. When the entire extended quantum collapses, it does so as a whole, vanishing from the field, like a drop removed from the ocean. Many have trouble accepting that, via collapse, the wavefunction instantaneously disappears from the locations it just moments ago actually occupied. A worry that is justified. But the alternatives — Many-Worlds or irreal interpretations like QBism or the Copenhagen — seem worse. In the Many-Worlds scenario, a world would manifest for every nail that it could interact with, and not to mention, in each of those worlds you'd still have someone wondering about collapse. Why did it happen here in my world? Also unsatisfactory are the irreal interpretations that hold that quantum theory says nothing of intrinsic reality at all, but only helps us to place in view the ways in which we are in causal contact with a "something" we cannot even begin to understand.

Anyways, all of this is to argue that collapse is physically actual. It is the dragon's bite that we see all the time and is more than throwing away a portion of an equation.

Quanta always act as units and collapse into atom-sized regions for the simple reason that only nuclei-bound electrons can accept their energy and use it to evolve. The "particulate" aspect of quanta is easily explained away when these considerations are taken into account.

Defining Decoherence

Quanta are delicate creatures and are easily affected. Even the slightest contact can affect a quantum spread out over great distances and cause them to collapse. Prior to this moment, when it remains unified and whole, a quantum propagating through space — or better, *as* its field — unperturbed, it is said to be *coherent*.

As a quantum evolves into a larger and larger superposition state, that it evolves to actually occupy "all the places" that it is, (or better, *is* in multiple states at once) its "states" are said to cohere. As we saw in the double slit, a coherent quantum is one that has evolved undisturbed and as such remains in a pure superposition state that will allow it to go through *both* slits. Or better again, there is coherence between the state of the quantum going through the upper slit and the state of it going through the lower slit. Put simply, quantum coherence captures the fact that superposition states obtain and are actualized. Even though we can see them evolve in this way, again, as interaction with the system will cause it to partially collapse, quanta really do evolve that way. We know this because repeated trials of the double slit will reveal the interference pattern telling of it having been in a superposition.

Now, given that, if one wishes to observe or acquire information about a particular quantum, one must do so by measuring it, by interacting with it, however slightly. There is no way to do this without coming into causal contact with it. To do so, one quantum must be used to measure another, and the moment of their "contact" will re-localize both, entangle them, and destroy the coherency of their *individual* superposition states. As such, the action of affecting a quantum's evolution to gain insight into this propagation is called decoherence, as it removes the "coherence" from the quantum's state. Effectively, decoherence marks the loss of the quantum's "quantumness." Decoherence does not necessarily *destroy* superpositions but actually *extends* them such that it now involves both systems.

The action of decoherence also reminds us of the fact that no system is an isolated system and that all of them, somehow or other, interact with their local environments. This is why, as we remarked earlier, decoherence commits itself to an observer-independent reality as environmental interactions constantly cause the localization of quantum states, *whether or not* people are around to observe them. Although not substantial enough to affect solitary quanta, the cosmic microwave background radiation (CMB) is enough to cause macroscopic bodies to quickly decohere. Decoherence marks the very way in which we can smoothly move between the quantum and classical "realms."

Decoherence is but another name for interaction as it always indicates an event or process where one quantum is affected by another quantum. The fundamental mechanism underlying decoherence is entanglement. In fact, the two are but different sides of one and the same coin where each side embodies certain characteristics and features. For instance, decoherence is all about the loss of superposition status while entanglement signifies the unifying merger of two or more quanta. When any two quanta interact, both processes take place.

To see all this more clearly, let's re-examine the double slit experiment but this time we shall finally add the "which-path" detector. However, consider first a "single" slit experiment using matter quanta — electrons — this time. If we shoot individual electrons through the slit, each will appear on the other side as a bullet-like impact, right behind the slit, signifying that they are particles. By retaining the marks made by each and after many trials a single band behind and in the shape of the slit will begin to come into focus.

This may be evidence for a particle ontology, but it can be more accurately explained with a collapsible wave picture for by it the same pattern will emerge. An electron as wave will exit a single slit and re-expands without interfering with itself — thus creating no interference pattern — as it makes its way toward the detection screen. In this picture, the same, solitary band will, after many trials, emerge.

DOUBLE-SLIT EXPERIMENT

PLANE WAVES

INTERFERENCE PATTERN

Now we put in place the second slit and observe the same results as the experiment done with single photons. Repeated trials will reveal the interference pattern. The electron, this time a wave of matter, comes through both slits, interferes with itself, and collapses into the detection screen.

But is it really coming through both? Is there a way to see which path it comes through?

A which-path detector is just as its namesake implies, a detector that can tell which path a quantum takes. When there is no which-path detector, the coherent quantum always goes through both slits, interacts with itself to produce an interference pattern, and collapses into an atom-sized region on the detection screen. A which-path detector serves the same function as the Geiger counter in Schrödinger's thought experiment but instead of registering radioactive decay, it tries to detect its namesake, that is, which-path the quantum takes.

Involving a which-path detector changes everything. To be useful, a which-path detector must extract information from the experimental quantum. To do so requires disturbing and interacting with it where the result will be a completely novel situation, an entangled superposition state that harbors correlations between the quantum and the detector.

In popular culture, a which-path detector is often made out to look like it is human dependant such that it's all about when we "decide to look" or is otherwise modeled in such a way that it relies on human agency for its actualization, but this is simply not the case. By saying the experiment changes when *we* decide to look somehow giving the impression that it is the human mind or "the observer" that interacts with the quanta coming through the slits. This is because when we do indeed decide to look the experimental results change, and a "particles only" result emerges.

However, the which-path detector has almost nothing to do with us. Save for the fact that we put it there, set up the apparatus, and decide when to turn it on... but that is true of experiments in general.

The true reason why the experimental results change when a which-path detector is added is that it introduces a *second* quantum system into the mix. Like every physical object to ever exist, a which-path detector is quantum mechanically constructed and uses quantum mechanical principles in its interaction dynamics. Therefore, to "see" which path an electron goes through a which-path detector *must* bounce a quantum of its own off of the traversing electron and recatch it. The *only possible way* to see "which path" an electron takes is to interact with it and as we know, any and every interaction always involves two things: decoherence and entanglement and it is here where things are fundamentally altered.

The detector bathes the slits in photons and awaits the recollection of one, a returning photon that will convey the fact that it bounced off of the electron at slit 1, or equally, bounced off of the electron at slit 2. Whatever happens, the detector will, but not yet, register that fact. When the detector's photon meets with the experimental electron the two will entangle and their respective states will decohere. This means that both quanta will arrive at the slits in coherent states – they will both extend over both slits simultaneously. Both the electron and photon will be meeting at both slits at once and it is here that they will randomly, for whatever incalculable reason, interact at *one or the other slit*. To meet, they must meet *at a place*, and by this obvious requirement, they simply cannot *interact* at both.

Here, their once-coherent, individual states will decohere causing a kind of partial collapse, where both will undergo an instantaneous re-configuration that recognizes and registers their interaction. At the very same time, this will entangle them such that their states will unify and *from here* they will evolve again as waves do. This is why a which-path detector removes all traces of the interference pattern. The decohering entanglement interaction resets the evolution such that both begin anew but from the slits themselves and restarting from here can only result in the singular band of impact points, one that sneakily and falsely speaks of a particle having just gone through the experiment.

But, having just left the slits we know not yet "which path" the electron took. Actually, that's a lie. We know all electrons always take both paths but when they are caused to readjust themselves at the slits by touching fuzzy photons, they are robbed of their ability to later interfere with themselves. It is only when they land upon the final detection screen that we may then correlate the result with that of the which-path detector. One that, due to entanglement, we can know, without doubt, will match.

In other words, the double slit with a which-path detector complicates the experiment by forcibly involving entanglement and its consequent decoherence. But once these are considered for what they are, the situation is seen as simple. At the slits, the electron and photon pair up to each form a unique subsystem of a larger, entangled superposition state. This entangled superposition state is called a mixture. It is a state wherein the system possesses actual but not yet made definite properties. That is, the electron and photon entangled at *either* slit 1 or 2, not both, but we cannot and will not know which until the electron's final collapse takes place.

An observant reader might ask, where has the superposition gone? The answer: in the *correlations between* the two subsystems. The "superposition" only reveals itself as a coherent relation between them and nothing more.

Decoherence is good for something else. As regards the quantum-to-classical transition, decoherence carries the weight of resolving the fuzzy, superposable world of the quantum into the vivid domain of our classical – solid bodies situated in space – experience. No one has ever seen a macroscopic object in a superposition of being both here and there at the same time and the reason is, once again, decoherence.

When two quantum systems interact, they each decohere the coherent superposition status of the other, entangling and thereby removing said status. Now, if every menial interaction entangles, localizes, and decoheres the objects that enter into a relation, then by it the entire enveloping environment serves the function of isolating and resolving the objects of macroscopic experience.

Obviously, the larger an object becomes the more it will be decohered, and thereby localized, by rogue photons. The fuzzy line at which this transition takes place is a calculable referent.

It is in this way that decoherence commits itself to a realist interpretation of quantum theory. Certainly, indefinitely many quantum processes, events, and interactions are taking place at this very moment. Planets none have ever laid eyes on are right now being saturated by their host star's EM field energy and some pebble on the far side of the moon keeps getting microscopically kicked around by cosmic rays. The universe is a wild place, constantly computing itself through informational exchanges that resolve its very whatness.

The Local-State Solution to the Problem of Definite Outcomes

Much earlier we saw how the measurement problem can be divided into multiple problems. The problem of irreversibility is solved trivially by entropy, but the key to resolving the definite outcome of Schrödinger's seemingly paradoxical zombie cat is to realize the feline is in a *mixture*, an *entangled* superposition state. It is not in the simple superposition state embodied by the photon of the Mach-Zehnder interferometer experiment but instead involves two entangled subsystems that come to harbor an instantaneous nonlocal connection.

Common understanding, and what Schrödinger thought himself, was that the evolution of his equation predicted an absurd, superposition state of affairs embodied metaphorically as a cat that is both alive and dead. It signifies a misunderstanding that has caused untold confusion and heated debate. But we may forgive him for "it is the privilege of the true genius, especially one who opens up a new path, to make great mistakes with impunity."[*]

He may have not misinterpreted his own thought experiment had he heeded the warning of the mathematical powerhouse John Von Neumann. For just three years prior to Schrödinger's thought experiment, Neumann pointed out that *a composite system of entangled quanta cannot have either of its parts in a superposition.* He showed that the composite parts that build up the entangled state cannot themselves be superposed but are instead in a statistical mixture or mixed state. This means that, while they each form a subsystem or part of the globally entangled measurement state, neither the cat nor the radioactive nucleus, *on their own*, can be in a superposition.

Now, could the composite system as a whole be in some sort of superposition? The answer again is no and for the very same reason: a composite system of entangled quanta cannot have *either of its parts* in a superposition. The cat is no zombie. This complicated situation is rather simple, it is a superposition of *correlations* between states. A situation that doesn't violate relativity or any of quantum theory's standard rules.

As it turns out, while the cat and quantum are in the box, they develop into a mixture of well-defined states, yet the outcome remains unrealized until someone opens the box to see. The cat is *either* alive and the Geiger counter doesn't register decay, *or* the cat is dead, and the Geiger counter registers decay. The system is in one or the other state, but it is not yet known which.

Again, an entangled superposition involves two systems: cat and Geiger counter, or in the case of the double slit, a which-path detector and electron.

---

[*] Voltaire.: *Le Siècle de Louis XIV.* (1751)

The entanglement that establishes itself between these two systems harbors an instantaneous nonlocal connection. A notion often overlooked but one that must be considered.

As we saw in the section on superposition, while traversing the Mach-Zehnder interferometer, a photon is in |this> and |that> state. The photon travels both paths; |this> state is superposed with |that> state and both obtain.

Now we change everything and involve a which-path detector whose unwanted effect is that it re-localizes the quantum such that it "only comes through" the slit the detector saw it come through. The effort to detect which slit the quantum goes through invariably affects the very thing we wished to measure. If we try to "dim the lights" so that the detector does not affect the electron, we reach a lower limit where we can no longer tell which path the quantum takes and at that moment the interference pattern returns.

So, to see "which path" it takes you we must interact with the electron. Which, via decoherence, partially collapses the electron wavefunction and resets it. Wherefrom this interaction point the electron will again begin to spread as a wave toward the detection screen. We now have an entangled measurement state correlated between the electron and the which-path detector's photon and we can rest assured the superposition pattern will vanish from the experiment as the decohering detector effect has certainly destroyed it.

The measurement state of Schrödinger's cat is an entangled superposition state where it is *either* in a definite state that correlates S1 with D1 *or* a definite state that correlates S2 with D2. We simply don't know which yet because the system has not yet collapsed. Although, when it does, we can be certain the correlations will obtain. The superposition of coming through |this> and |that> slit – of the cat being alive and dead – is destroyed by the interaction with the which-path detector and the superposition evolves to now correlate the which-path detector and the final result.

There is no zombie cat and never was. Simply speaking, Schrödinger's original examination was flawed. Again, the quantum plus cat is in |this> *or* |that> state but prior to opening the box (measuring the state) is indetermined. And what is superposed – read instantly and nonlocally connected across the experiment – is the relationship between the quantum and its detector. The correlations between the subsystems are what is superposed. The entanglement caused by the interaction limits the state of both in such a way that neither individual quantum is in a pure or coherent and therefore superposable state.

To recap; the measurement problem is solved trivially when one realizes that as the superposition states of the subsystems become entangled and will thus, forever after, share an instantaneous interconnection, decoherence will, at the same time, evolve the superposition to involve the two. The result is an *entangled* superposition state. What is actually superposed is not a cat or radioactive atom or even the composite system, but rather, the *correlations between* the two subsystems. With a detector involved, decoherence causes the measurement state to evolve into a *mixture* of being in a well-defined yet — until final collapse — unrealized state. When the measurement is completed, due to the nonlocal effect of entangled systems, one will find a direct correlation with the detector, *even if* that detector happens to be on the other side of the universe.

All this was originally seen and put forward by Joseph Jausch in 1968 and has since been rediscovered by many. The resolution is simple and falls in line with the standard quantum mechanical rules although it adds the further commitment to collapse as a real physical process. Art Hobson has dubbed it the "local-state solution" because one must examine separately the local-state result of each entangled sub-ensemble. In other words, one must look first at the slit-detector and read its result and later look to the detection screen to see where the electron collapsed. Entanglement will always ensure the proper correlation.

Should the reader wish to examine the local-state solution to the problem of definite outcomes in greater detail, it has received its greatest explanation from Art Hobson in his swan song Tales of the Quantum. Not only has Hobson put a name to it but has also provided new arguments for it based on special relativity's ban on superluminal signaling. Should you require a far more detailed explanation please see his book.

.

Part III

The Mindfield

> "The pure present is an ungraspable advance of the past devouring the future. In truth, all sensation is already memory." - Henri Bergson

CHAPTER XVIII

The Field Whose Property is Awareness

Remember way back when I mentioned QFT doesn't claim knowledge as to "the what" a quantum *even is*, save those three clues that let us know that as a "thing" it harbors and outstrips its wave and particle precursors. Nor does it commit itself to a particular ontology and allows its many "interpretations" to stand as metaphysical theories all on their own.

Well, let us add another. What if a quantum system could beget such entangled, structural complexity and internal informational processing that it could become a pocket of sentience, a continuously variable yet discretely differentiable unit rippling through a unified field of awareness? What if by involving every quantum field in its constitution, the quantum system in question begins to tap into an always-already-there but not as-of-yet activated field? A field whose property is not gravity, electromagnetism, or boson-glue, but *awareness*. This is just what I wish to argue and will claim that the most primitive quantum system that achieves this is the living cell. Furthermore, when cells build with and upon one another they entangle further still to become larger-order living structures, tissues, organs, and organisms to eventually end up with the most complex, unified quantum system known to us; mainly, ourselves — the human being.

At long last, we are finally able to attempt the construction of a model of consciousness based on quantum-theoretical concepts and commit ourselves to an ontology as regards it. It is one that holds to the reality of a quantum system as defined by its wavefunction and understands interactions as grounding and necessarily self-defining. What is meant by that will be examined in what follows. It also recognizes information as an immaterial structural scaffolding that helps to define the "isness" of all objects.

The primary concept of the ontology here developed is called the mindfield and it defines consciousness as an *actualizable, intrinsic property of the non-void-void-vacuum quantum foam.* It holds that "mind" is a *realizable characteristic* of space-as-nothingness that is — like all quantum fields — always-already *there*. Except, whereas fundamental quantum fields are actual *determinable properties* of the quantum foam, consciousness is imbibed within it as a potentiality and is only "reached" or "activated" when certain coherent portions of said foam become sufficiently complex. In that, the mindfield is a composite field that includes many, if not all of the fundamental quantum fields in its constitution (and probably several as-of-yet not identified others).

To "reach" and thereby activate the mindfield, a maximally entangled "quantum system" of sufficient size and complexity must first obtain. When it does so, just as the Higgs mechanism causes electrons to divert some of their energy into warping the Higgs field, the now-organic system will actualize novel properties of Nature itself and thus actuate "awareness" — however primitively — as one of those previously inert qualities.

Certainly, there are, of course, many unrealized characteristics of the quantum foam but the particular one with which we are concerned is the field whose property is *awareness* itself: the mindfield. By its active realization, primitive awareness will instantiate itself and its action will be seen as a new kind of "force" known intimately as will. Primordial awareness will become an innovative gravitas that will, by creative evolution, usher into the universe indefinitely many novel beings. Eventually — as is realized by *every single human being* — nature will invert its eye and the universe will come to know and reflect upon *itself*.

But how do we get here? What is the primary step necessary towards establishing primitive mindedness?

On this planet, the most basic system that has achieved a level of amplified complexity capable of stimulating the mindfield into an adumbrative being is the primal prokaryotic *cell*. A cell is the most elementary unit capable of carrying out the basic process of life and forms the "quantum" of the mindfield, its minimal agent. The cell is the basic unit of all living things. Some of which are made up of single cells (amoeba and bacteria) or multiple cells (plants and people).

Despite its already incredible complexity, the primitive cell possesses the lower-limit structural energetic organization and functional informational intricacy such that can trigger and actualize the mindfield thus becoming primitively "aware." Strange-loopily, it signifies an actualization that bestows upon the system an awareness of itself.

The cell *knows* nothing but instead "senses" in a highly rudimentary manner. Yes, even the simplest cell is in some sense "aware," but the depth of degree by which it is, is of course, quite minimal. We can infer this by observing the actions of certain microorganisms — like those equipped with flagella — as they demonstrate the ability to actualize *purposeful behavior*.

Attentive researchers have watched as these tiny lifeforms switch between chaotic and deterministic behavior. In their search for nutrients, they will flail their flagella chaotically and then ride the wave set up by it. Every so often, as they tumble through a solution, they will "sense" a nutrient grade and will begin flailing their flagella coherently to move toward the reward. This illustrates that certain cells enact behavioral choices. Choices made available to them by their primitive ability to "sense."

Bright readers will also understand this dynamic as an automatic physiological reaction, the most basic psycho-behavioral imperative: the fight or flight reflex. That it happens automatically, like most of an organism's internal processing, does not also mean the movement lacks awareness. Indeed, awareness is a by-product and corollary of that very drive.

Cells are not intelligent, but this does not mean that they are also devoid of minimal sentience and elemental awareness. Just because they lack sense organs does not mean the ways in which they process information also lack infinitesimal *sensation* regarding said exchanges. As we saw with human sense organs as channels of information exchange, now we claim that *any* channel of information exchange may mark a nexus wherewith consciousness may establish itself, however minimally.

The biological sciences have revealed the cell to be an autonomous, self-constituting, autopoietic system. Truly, a complex system, it can be seen to involve nearly every quantum field in its composition up to and including the gravitational field while its internal structuring is incredibly complex. Again, although the unimaginably complicated chemical constitution of a cell is incredibly complex, many believe that there is no intelligence there. That the internal bits of a cell simply undergo mindless chemical reactions, but it may just *look* this way to us. It could very well be that the teleological action of a cell's internal apparatus may be the first instantiation of a primordial and very primitive will.

Not only does the cell have structure and a high degree of order, but as the primal unit of Life, it performs many fascinating functions. Cells respond to stimuli, they grow, develop, and reproduce, they use energy to power themselves and evolutionarily adapt to changing environments all while maintaining homeostasis. Homeostasis is that autopoietic, internal, self-regulating process by which a living organism remains a stable system while responding to changing, external, environmental variables.

As with every physical object, the description of a cell as a living organism is captured by its wavefunction, a roster of its ontological, categorical properties. It has a weight, a mass-density, and a measure of intrinsic informational coherence. Its internal subsystems are quantum mechanically entangled such that changes undergone by spatially separate parts will be immediately registered by the system as a whole. To be sure, some of its internal components may be entangled to a greater or lesser degree but on the whole, the cell will be seen as a single quantum system.

Every feature of a wavefunction is realized by its own oscillatory, energetic extension into that field's property. A cell actualizes itself as a portion of quark and electric field energy, gravitational potential energy, and cloaks itself under virtual EM field energy. A register of its properties, each of these are "dimensions" that the total energy that the cell as quantum actualizes. More than this though, other "dimensions of being" are actualized by the computational processing of information that begets its internal organization. The capacity to "sense" nutrient gradients, for instance, becomes but another extension of its wavefunction. A place wherewith its fuzzily extended being reaches out to "touch," — that is, "sense" — its environment.

As we know, quanta transcend both wave and particle concepts. Although they are more akin to waves than particles, they may be better understood simply as *wells of influence* whose dynamics are like the two-way street suggestive of gravity such that their action is to *pull-in* while *reaching-out*. Quanta may be better understood not as vibrating, oscillating objects, but instead as nebulous and fuzzy spheres or shapes of a field's property and potential. In this way, an organism may always-already be "touching" other objects that inhabit its immediate surroundings. Maybe, by seeing what is "out there," I am really and truly in touch with it, as the image of my vision is made up by the exchange of information along that channel. And that what is "out there" is really "in me." Indeed, it is both.

Now, as for the wavefunctions that characterize creatures, the "dimensions" it comes to activate signify new *ways* that it may sense. A "direction" that it may come to know. Seen in this light, Hilbert space is not merely an abstraction, mathematical structure, or logico-architectural space, but is instead actual, *real*, and indicative of the true nature of space and time. Atop of this, maybe reality is not only infinite-dimensional physically, but infinite dimensional imaginatively with regards to consciousness, and every sense it is able to achieve forms but one of those dimensions, a "direction" or "space" that a particular wavefunction may oscillate in and actuate.

This is the image I have in mind with regard to human-grade consciousness. The objects that we are in touch with are revealed by our *relation* to them, in the horizontal "modes of givenness" that beget presentation. These are the channels of information exchange where its processing is *known* as a sense. There are real and imaginary dimensions that wavefunctions extend into. The "real" dimensions are the most physically vivid as they define the extension of Being, namely of matter and its dynamics. Those are of course classical space and time. While the "imaginary" dimensions are epistemically vivid, like the judicative domain of reason, or the playful realm of imagination.

Indefinitely many dimensions of Being exist. Broadly speaking, the real is known perceptually while the imaginary is known conceptually.

But what of the loss of quantumness with regard to the single cell? Why are they never seen in a superposition of being both here *and* there? Simple, as a mesoscopic object, its superposability is quickly decohered by its being embedded within and enveloped by an aqueous environment. In other words, objects of this size are constantly localized by the surrounding sea of interacting energy. Even if a single cell could be sent through a double-slit in vacuo, cosmic rays and unremovable photons that form the cosmic microwave background radiation would quickly decohere any such mesoscopic object.

Through decoherence, every macroscopic, coherent object's environment will serve the function of localizing it, effectively removing the system in question's quantumness. Every interaction selects the "whereness" of the object while defining a new relation.

The mindfield is conceptualized to be, like Hilbert space, infinite-dimensional. And the dimensions that a certain quantum actualizes are the ones relevant to it and its local environment.

A central point regarding this kind of oceanic ontology can be seen in the relationship between a field and its quantum. That every electron is an excitation of one and the same underlying field speaks to the inherent unity of all electrons. Likewise, every expression of the mindfield is unified in a similar fashion.

Anything and everything in existence, no matter their difference in distinction, is an expression of one and the same thing. That is, everything that is is an expression of the self-same quantum foam. Even though all things have autonomy and subsist on their own, the Many are unified when we realize it is an expression of the One.

This mechanism can be applied to every categorical species in existence. Every person, no matter how unique or complicated their history, is an expression of the self-same mindfield and it is here where we can recognize our shared, inherent fundamentality. Only this quantum mechanical ontology that defines the beingness of objects through their relations and self-constitution is able to rectify the paradox regarding the Many and the One. Indeed, everyone is an autonomous being and is capable of self-determination. At the same time, we are all expressions of one and the same underlying thing, the field of consciousness itself, and by its realization reality achieves reality valuation.

Consciousness is a unified field that we each in our own unique way, actuate and are. That which constitutes me constitutes you. Most deeply, we are the same, merely different expressions, of the same immortal energy. I am you and you are I and together we are Divinity itself.

Although every electron that inhabits the electric field lacks *primitive thisness*˙ and is thereby indistinguishable from their brethren, they are each an expression of the same underlying thing. Each is a quantum of the electric field. In just this way, we are, each of us, separate, but at the same time, expressions of one and the same thing. This grants the mystical insight into the unity of mind and nature new evidence and is best expressed in the timeless texts of India, the Upanishads: "I am all this creation collectively, and besides me, there exists no other being."

What individuates us is the great chain of being, a historical line populated by moments of interaction. By bundling structural properties and dynamical functions onto a cell, and then building upon that complexity platform in a myriad of ways we finally reach the intricacy of animals and the exquisite form of consciousness they activate. As people, what we know personally are the avenues we've explored from our restricted, individual vantage points in space and time and the incalculable moments of information exchange, both internal and external.

Communication, Sociological Entanglement, and the Lived-Body as Reference Frame

A quantum wavefunction is a superposed multiplicity of actualized degrees of freedom. This, like the possible emergence of spacetime via this mechanism, shows that a single quantum possesses a depth of degree insanely complex and is indeed many things at once and is pure but limited potential. It is simultaneously actual and can only be what it already is. Its evolutionary possibilities are not infinite but are historically restricted and sets the limits on what it is and can become.

That quanta aggregate – that they absorb others into themselves such that what they absorb lose their individuality to become the whole (basically, they sum by superposing into one) – allows us to see how quantum systems can become so internally complex all while remaining in a state of wavefunction

---

˙ This helpful distinction comes from Paul Teller and captures the philosophical idea of 'haecceity.' This refers to a property or quality of a thing that, by virtue of which, actualizes its uniqueness or allows it to be described as this or that *one*. Again, free, elementary quanta *do not* possess any attribute, property, or quality that could, even in principle, distinguish them from others. For that, time and indefinitely many entangling interactions are required.

homeostasis. Once aggregated by maximal entanglement, previously individual quanta lose themselves and contribute only to the organization of the whole.

That a wavefunction is a superposed multiplicity of various degrees of freedom is similar to the situation that presents itself as conscious awareness. Simply think of all the things you are at once: a body, a mind, a movement, a feeling, a sound, a sight, a symphony, a seeing, a sensation, a sentience.

"You" are not just a dizzyingly complex 10 to the power of 56 entangled quanta — a living wavefunction — by association, you are the whole unified field itself. The two identities are one and the same. Remember Auyang's distinction of a field as a whole with differentiable parts. Two electrons in the same field are identical, and indistinguishable. Yet there is a distinction. A difference. Mainly there are two of them. They are not the same, yet they are identical, each an expression of its underlying field. One could swap their places and there would be no way to tell which is which, unless of course, they harbored a lived history. That is what makes us all the same yet incredibly astronomically different and unique.

Now, with regards to the field of awareness itself, when two separate people — two minds — come to occupy the same environment, although each specifies a field configuration concerning their particular perspective and remains "open" along channels of immanent possibility, both activate and know the same field. Although each will know an awareness tinted by their historicity and bodily being, there is an opportunity *to know one and the same thing*. Not one thing is known in two ways, but one thing is known. A portion of each's consciousness "overlaps" and again, they know *literally* the *same thing*. It is not the case that each knows their own version of the thing, as this would multiply it ontologically. Said differently, when two cerebral holographic hallucinations overlap, their maps superpose and present an opportunity for each to know something exactly as the other does for if they both grasp it, they both grasp the same thing.

For completeness's sake, consider that "something" (a revelation in a movie, a promise made, witnessed act, or what have you) is a unit or piece of information, a bit. Now, when two minds entangle, they each walk away with a carbon copy of *that bit*. Each knows the same bit. Not two bits each known to one another separately but one bit *shared* by them collectively. Equally, we could say that the entangling of two bits *generates* — *creatio ex nihilo* — a third bit that is known to both.

It is this interpersonal nexus — established between two or more people — that *generates* human social reality where social constructs are like that third bit.

Something is social if, and only if, it is the result of two (or more) minds interacting with each other knowingly. The nexus between them forms a novel kind of space wherein a novel type of object may appear. Once it appears to it becomes established; essentially, becomes a thing. Commonly demeaned as a "social construct," this now-actual object has been etched into the historicity of the universe. Its genesis marks the moment in time where it came *to be,* no matter how fleetingly. Even though the existential value of social objects is incumbent upon the minds of the people who imagine them, this does not reduce them to unreality.

Social constructs are as real as objects get for us. Again, reality is, for us, that sliver of the Infinite that we are in touch with. A portion of which encounter and another we create. Matter is the part we encounter; mind is the power that allows us to create. Both equally exist. That social constructs are based upon or acquire their being via our imaginative faculties does not mean they are not part of this world or are unreal. Certainly, the Infinite has endless space for them.

As with the genesis of new information by quantum entanglement social objects come to be at the intersection of communicating mind such that communication is directly perceiving a higher order. A higher order that is but another dimension of the Infinite.

A rudimentary example of the difference between instinct and language-expressed intellect can be seen in the recently discovered news that great white sharks are afraid of killer whales. Because orcas possess social behaviors that sharks do not, the orcas are the superior predators, and great whites' flea them in fear. That the orcas are the superior predator is due to the fact that they can *communicate.* They use cryptic language that enables them to "see" something that the sharks cannot. In a word, orcas possess a consciousness of a certain kind of higher order, mainly, a sociological one. By communicating, a novel domain of Being — pertinent only to them and their language-games, but nevertheless equally actual and real — actualizes itself and becomes part of Being, part of *what is.*

Every human being — every living creature burdened by sentience — can be considered an inertial frame of reference. The consciousness that you are now as you are reading this forms a coherent and well-defined frame of reference. As we know, any organism is a well-defined, inertial, physical system characterized as a wavefunction.

The fragment of the field of consciousness that you are — that reaches into and out from you— forms a well of influence in the holographic mindfield. People can naturally tell when they are in the awareness of another. When two minds encounter one another in the same field, their awareness sums such that they both share an awareness of what is transpiring locally. Internally, their monologues will differ significantly, but externally, they *share* an awareness. Given the freedom of the will, either conscious agent can direct their intentionality toward an object of their own mind's choosing, but there will always remain an *option* for placing their awareness on what can be shared between them. This need not always be something external to the two.

In the same vein, one may enter a room where they can sense the tension of a recently transpired argument between a couple or likewise, can read a room and sense the general mood of a crowd.

In many instances, something internal to both can be shared as a single knowable phenomenon. Meaning in language, mutual understanding, and so on. True unity of internal states between minds is most powerfully explicated by transcendent lovemaking.

As a frame of reference, an invisible "boundary" exists between lived bodies. A line where the difference between the frames must transform to allow continuity and coherency to the evolution of the universe. The transformation that takes place at the periphery of an intermingling of multiple minds is called a Lorentz transformation. At the nexus where they meet, not only do they interact, but between them, the Lorentz transformation keeps the universe's evolution coherent. Between moving, interactive systems the universe will shift simultaneity, slow time, and contract space in such a way that the proper time interval will remain conserved.

Concerning how fast creatures move relative to one another, these effects are inconsequential. If it weren't for special relativity and its spacetime symmetries, we'd never know of them. But they are nevertheless real (de)formations in the structure of reality. And if they take place between inert physical systems, there is no reason they wouldn't between minds.

We get to know people by processing the information *exchanged* by them along two separate channels, one spatial as it is their body image as seen in the visual image but better still through intuition by perceiving their vocal image as it moves through time. Along this line it can be further analyzed into its meaningful content and where particular attention is paid to nuance and inflection.

A pattern integrity is a larger order structure that — as a whole — transcends its parts. A band has this structure. As its members are replaceable, within reason, the band persists even though it's held up by the intentionality of the members. A person has this structure, replace every cell and the person will remain. A wavefunction has this structure.

When two quantum systems come together and entangle, each retains the same bit of information regarding the interaction but each in itself. This is why maximally entangled quantum systems, like lived bodies, are each at the same time their own "thing" while also, at least partly, being the other.

Not totally, but people are localized — that is, defined by — other people. The self that I am is grounded by its being perceived by others. The memories we together make ground us and help to form us. The person that I am is not only contained within me, but I am also in the hearts of my family and the minds of my friends, and now, as the author of these words, whether you like it or not, part of me lives on in you.

On Death / Further Dimensions Opened to us by Thought

Possibly his best passage, with Death on his mind, Shakespeare tells us that "Time hath a wallet at his back, wherein he puts alms for oblivion," subtly reminding us that annihilation awaits us all. That no one, no matter what, can escape its unending march. Death is the extinguishing of consciousness, the act-movement where we come to no longer become. It marks the moment where our consciousness is lost, returning to, and settling into the sea of awareness from which it initially arose.

Death and the knowledge of it as everyone's destiny can be seen as either a blessing or curse. Knowing that someday we will no longer be, may serve to encourage or hinder our own self-expression. For some, it allows them to live with passion, for others, to live in fear. But either way you look at it, knowing of death is a way in which we know about the world and our relation to it.

As I have argued, the various *ways* that conscious systems sense indicate the "dimensions" in which it extends and operates. A composite, psychological sense, by coupling our capacity to think with our ability to convey information through language opens to our awareness many previously unknown avenues such as the dimension of our being that is toward death. A spineless Nazi[*] first pointed this out — that there is a portion of our mind that extends, is oriented to, and is structured by the *knowledge* of our inevitable annihilation.

If we encounter the emancipating knowledge gifted to the mystic, to those with the eyes to see rightly, we need not fear death as we are truly immortal. Instead, we lose only our bodily form by returning it to the timeless Substance that spawned it.

To be sure, it is not the moment of death's occurring that causes us existential anxiety but rather the many ways in which can *imagine* it happening. Being burned to death or drowning marks miserable ways to go. Being tortured to death yet something entirely more horrific. Regardless,

---

[*] I am here referring to the unapologetic, should-be-forgotten, piece of shit, Martin Heidegger.

we *know* with absolute certainty that death ever approaches and, in this way, forms a part of what makes us *us*.

Before its advent, consciousness knows and illuminates the fact of death but is itself never known as its transpiring marks its extinguishment. Without bodily form to transact information from the field to the field there will no longer be the light of one's consciousness.

Now, much has been said regarding the mindfield. Before leaving you, I should like to accentuate two facts regarding its inherent nature. That it is immaterial, and structurally, achieves its essence by exploiting information as a relating "substance."

> "The question of *creatio ex nihilo* reduces to explaining how some information arises out of nothing. Information is the *only* concept that we currently have that can explain its own origin." — Vlatko Vedral

CHAPTER XIX

Creatio Ex Nihilo | Efflorescent Complexity and the (Bi)Unitary Nature of Information

The epoch in which we live is the age of information. A time defined by computers and the many digital spaces that they construct. The internet is its penultimate achievement. A cyberspace that near instantaneously interconnects the whole planet in real-time. It is the result of countless considerate minds, and it runs on the progressive computation of information.

A topic about which we have said much, tacitly understanding its meaning, but we have not yet defined it. A difficult task to be sure, given its variated nature as it seems to sway between the ontic/epistemic divide. Is information a real, tangible "thing," or does it represent a degree of knowledge? As it turns out, it's both and probably much more. Physically, this "quantifiable measure of communication" reveals itself when its mathematized.

The mathematician responsible for such a feat was Claude Shannon and as such is credited with the origin of information theory. The philosophical groundwork, however, had been previously laid out by, to name but a few, Aristotle, Boole, Russell, and Turing. Without their — and many unnamed others — previous contributions to the discipline of logic, the information age may have never dawned upon us. This groundwork is what allowed Shannon to reach his quantitative conclusions in his 1948 paper *A Mathematical Theory of Communication*. In it, he would define the bit as the fundamental unit of information. A bit being shorthand for "binary digit." A bit of information represents the possibility of a dichotomy, basically, an — and/or, on/off, up/down, this/that, 0/1 — binary option. Classically, bits are in one or other state and flip back and forth as they interact. It is a one that can become a zero and vice versa.

Shannon would go on to reinterpret the thermodynamic notion of entropy in terms of information. In fact, the two share essentially the same equation. It reveals the efflorescent linearity of information exchanges. That every instant

Creatio Ex Nihilo | Efflorescent Complexity and the (Bi)Unitary Nature of Information

the universe computes itself it becomes increasingly more complex. Thanks to Shannon, the quantifiable result ended up being expressed by the credo that "information is physical." As is the case with quantum theory, the classical version fails.

By respecting quantum states, the bit needs to upgrade itself to become a qubit. This "quantum bit" is capable of existing in a superposition state. As it turns out, our friend "spin" becomes the primary referent.

Due to superposition, spin "rolls" in every possible direction but as to its quantized disposition, the symmetry principles of spacetime dictate that nature must "pick out" a direction concerning it such that it can only ever be defined as a $|this\rangle$ or $|that\rangle$ quantity, thus revealing spin as *the* primordial dichotomy, disclosing it to be the most basic qubit.

As we saw in the section on entanglement, in his *Programming the Universe*, Seth Lloyd remarked that "quantum mechanics [can] create information *out of nothing*."˙ Due to the entanglement of two qubits, a third qubit of information is generated that was not present in the initial two. The entropy of the whole is raised, seemingly from nowhere, by the entanglement unification of its subsystem parts.

It is time now to speculate as to how this may have initially taken place or been the case with respect to the dawn of time itself. I should like to consider the genesis of novel information as one possible facet and the random manufacture of a "rogue wave" as another.

The latter idea was first explored by David Bohm in his timeless *Wholeness and the Implicate Order*. There, Bohm imagines the incessant effervescent bubbling of the quantum foam to produce periodic waveforms whose summation *may* result in an outstanding waveform. A waveform which, as recently confirmed by statistical means, superpose to produce a monumental "rogue wave." A "rogue wave" that is of such unexpectable proportion that it causes a tear in the fabric of spacetime and causes it to *inflate* at a terrifying rate.

However, the former idea may make more sense as it is the establishment of information as a content of relation that begets the genesis of something from nothing — of *creatio ex nihilo*, as the ancient Greeks may have said.

Speculatively, the quantum foam is thought to predate the universe. It is a structure that imbibes possibility as its being but is itself not yet a thing. In his *Something Deeply Hidden*, physicist Sean Carroll examines how the universe may define itself through the entanglement interactions of its most

---

˙ Lloyd, S.: *Programming the Universe*. (2006)

substanceless and minuscule "parts." A movement to be defined as the unification of a void's entangled degrees of freedom.

You see, like the non-existent "springs" that allow us to envision quantum mechanical harmonic oscillators, neither do "degrees of freedom" exist in that very same space. Wait, I take that back. Both the oscillators and the degrees of freedom that belong to them exist in that location... but not as we have defined them. They are more accurately, both of them, mathematical abstractions and in the physical space where they are modeled to reside, they constitute *possible ways of being*. They not yet *are* but nevertheless indicate what is possible. We may say that they are *potentially* actual.

The ontology so revealed is one of unified immaterial information. Holographically, this may be the case as information need only recognize itself — that is, relate to itself — to structure the baseline actuality of total existence. Furthermore, information is easily understood as an immaterial structural facet of the quantum foam, and it marks the incalculable interactions between its countless parts.

To recap: what we have come to discover is that the world as we know it achieves actuality as a set of information-exchanging, operator-valued, quantum fields. The quantum foam as carrier of these fields allows them to interpenetrate one another and through their composite interaction access novel fields and thereby activate previously inert qualities of said vacuum. As with all fields, these bare qualities are timeless and eternal. Consciousness, or simply awareness, is just such a quality, one in which we live intimately.

As we know, quark, gluon, EM, psi, and the Higgs, all come together to structure a globally transforming, universal matter-field. A unit of energy that actualizes it is the field's resonant quantum.

A field is a continuous whole, whose deviating amplitudes as shapes are its objects. The fields are indicative of the void itself and their warpages are their resonant quanta. An unfathomable number of quanta occupy and are the same field and thereby share as common this common-to-all ground floor.

Although intrinsically immaterial, quanta couple to form the bricklike building blocks of the universe and as they come together, they entangle and localize themselves. Captured by the nuclei of atoms, electrons, and the electrochemical forces between them allow for aggregates of quanta to entangle and sum and grow to become unified wholes that form and are formed by deeper field magnitudes. For instance, a proton is a semi-localized, as-constant-as-can-be-phase, energy amplitude of the proton field. It is a movement and well of influence that warps and affects its field in such a way

Creatio Ex Nihilo | Efflorescent Complexity and the (Bi)Unitary Nature of Information

that allows itself to make contact and interact with other quanta, other amplitudal shapes, present in the field and those it is coupled to.

As we know, when a quantum of field energy encounters another, they exert forces on one another and entangle. Essentially, the two become one. As this process continues with regard a single system, we arrive at aggregates of quanta that are so immensely entangled and decohered by their environments that they behave classically and appear as solid, isolatable bodies in space. In a word, classically.

Every object so defined is surrounded by a shield of electrical energy that communicates with other objects by exchanging virtual photons through the intermediary EM field. Again, *every* single isolatable, solitary object achieves its autonomy by being electro-chemically bonded to itself. Within itself, an object may attain a lesser or greater degree of entanglement. For instance, although a rock may be a whole object describable by a single wavefunction, it may not be wholly entangled. Living systems, however, *must* be maximally entangled.

Now, we must keep in mind that a sufficiently entangled aggregate of quanta can be considered a single quantum system, characterized as a wave, described by a wavefunction. This unit of being is true of all existent objects – including your "self."

Imagine an atom and the standing wave of the electron cloud that pulsates about it. This lone atom is described by a single wavefunction and codifies a solitary deviation in the universal matter-field. When two atoms come together, they will entangle and form an even greater deviation of the field's expectation value. This will go on and on and on until we arrive at a macroscopic object.

Picture extended material objects as unified standing-waves of such coherently entangled and tightly bound quanta that the intensity of the matter-fields deviation at said location localizes the object itself. Matter is only a blob of highly compact energy that incessantly vibrates. This is of course, why you can shatter a wine glass by sounding its resonant frequency.

Also imagine that within the animate matter that begets Life, the complexity of its wavefunction as a pulsating standing wave may reach into novel fields hitherto unknown and not-yet-classified. Remember also that all standing waves are themselves *wells of influence*. Like the bosom of Khaos, a well of influence is a bidirectional movement that simultaneously *reaches out* from itself to *pull in*. "Reaching out" signifies a self-to-other movement while its consequent "pulling in" denotes its opposite, an other-to-self development. Such is the case with every "horizontal," presentational mode of givenness.

Creatio Ex Nihilo / Efflorescent Complexity and the (Bi)Unitary Nature of Information

In just this way, every coherent quantum system — whether inert molecular structure revealing itself as a gem or rock, to every human being self-determining itself as a person, or to every eukaryotic cell circumscribing itself as a self-organizing organism — represents and is indicative of an individual object; whatsoever those objects relations to others may be.

As organisms evolve and grow, so too does the technology the use to interact with their surroundings. A result is that specific senses organs establish themselves and come to mark *vivid* ways in which reality can be known. Indeed, the acquisition of a particular sense organ may mark an efficient coupling to whatever field it so reveals as a dimension of being and known as a perceivable sense.

Consider how the chemoreceptors on your tongue open the door to the dimension of taste and flavor. These same cells are present on the skin of earthworms, and I don't think it is a stretch to imagine that they possess a rich inner life although the sentience may only equate too an amalgamation of different "flavors" of dirt. Certainly, worms are made "aware" of the soil's nutrient density by this very function and behave accordingly.

But we are getting off-track and must turn now to one final consideration implied by quantum theory as it marks a novel architectonic structure with regards the nature of the universe... as a hologram.

"What we perceive through the senses as empty space is actually the plenum, which is the ground for the existence of everything, including ourselves... This plenum is no longer to be conceived through the idea of a simple material medium, such as an ether. Rather, one is to begin with the **holomovement**, in which there is an immense 'sea' of energy. This sea is to be understood in terms of a multidimensional implicate order, while the entire universe of matter as we generally observe it is to be treated as a comparatively small pattern of excitation. This excitation pattern is relatively autonomous and gives rise to approximately recurrent, stable, and separable projections into a three-dimensional explicate order of manifestation, which is more or less equivalent to that of space as we commonly experience it." - David Bohm

CHAPTER XX

Holography | To Write the Whole

Ultimately, we have seen that quantum theory allows us to denote, with clarifying ontological precision, just what exactly it means to be an object. A description that recognizes the necessity of its relations, especially, as is the case with environmental monitoring that begets decoherence. In other words, the linear "superposability" of quantum theory grants us a unique way to isolate "objects" within other objects — to see how an object as a whole also forms a part of a larger whole, and so on, ad indefinitum. When quanta aggregate into new wavefunctions that assimilate so totally that they remain a single object.

Now, to better understand the notion of wholeness with "separable" parts, we need to analyze the nature of a hologram. A 'hologram' is a complex object that reveals itself as a three-dimensional image encoded on a two-dimensional surface while 'holography' is defined as the process of wavefront reconstruction. The word itself means "to write the whole" as it comes from the ancient Greek *holos* meaning "whole" and gram meaning 'that which is written.'

Its basic structure is that of a higher-order image or object — a single step-up in dimensionality —that is constituted by a lower-order information store. This higher-order object or image is called the "bulk" and the information for its actualization is stored on the hologram's lower-level "boundary" and is everywhere the same for the image so produced.

Because of this, a hologram can be divided into multiple copies of itself with no loss in information content such that if it is cut into many pieces, each one will project the entire image, but as if viewed from a smaller subset of

Holography / To Write the Whole

angles. That is, when this information is re-sutured together by an appropriate reconstructive wave. Further to this, because every "part" of a hologram contains within it the information to reconstitute the whole, the same photographic plate can embed multiple — vividly different — three-dimensional images.

But we must remember, the information regarding the hologram is not where we think it is (i.e., in the bulk) but is instead "on" the boundary.

To see how this is possible, let's consider the process by which one creates a hologram, and then we shall apply it to the theory of consciousness as a multidimensional field.

To create a hologram, we need an experimental setup that somewhat resembles a Mach-Zehnder interferometer. We begin by shining a specific frequency laser into a beam splitter. The light that is transmitted is called a "reference" wave and it moves directly toward our photodetector, which this time is a plate of photographic film. A sheet of gelatinous paper will record the interference pattern that the light will carve into it. The light that is reflected from the beam-splitter is called the "object" wave, for it moves toward and is reflected by whatever object we are creating the hologram of—in this case, my logo.

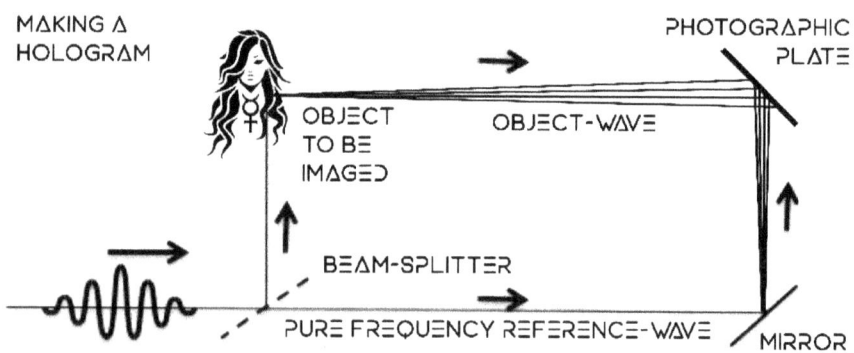

The two wavefronts then reconvene at the plate where the information of their interference pattern is etched into it and the logo may or may not appear.

As an aside, in principle, one could create said hologram by using single photons! Although, because of how they collapse when interacting with the emulsion, it would take quite some time and many photons to build up the scene. "We performed a relatively simple experiment to measure and view something incredibly difficult to observe: the shape of wavefronts of a single photon."[*]

We have created a hologram, but it is — in a sense — without being. The information is there, carved into the plate but without the proper light is incoherent. If one were to look at the plate with light harboring multiple frequencies, one would see only a diffuse mess. But when light is shone on the plate that is made of the same pure frequency as the reference wave used to create it, the hologram comes to life. The recorded interference pattern (re)diffracts the light such that the logo steps out of the plate. Thus, for the hologram to subsist properly, it must be brought back into being by a frequency-specific "reconstructive" wave. Only then does the image stand out as truly three-dimensional, possessing depth of field, showing parallax, and will change realistically when the perspectival viewing angle is altered.

The main takeaway is that we have created a three-dimensional image by storing its information on a lower-dimensional boundary. Neat trick.

Before moving on, let's consider in greater detail how a hologram "writes the whole." First, consider how the light coming from the beam-splitter bathes our object to reflect from every one of its crevasses, surfaces, and features. From every point the light reflects off the logo, it will expand spherically such that when the wavefronts reach the photographic emulsion, they will cover and interact with the entire plate. As such, the information of every definable point of the structure arrives at and is recorded in every definable point on the plate.

Now, if we cut the hologram in half, we are not left with the upper part of the logo on one and its lower half on the other. No, what we now have are two complete logo holograms. We can continue cutting and we will always end up with complete holograms eventually arriving at a lower limit. The main takeaway here is that *every part contains within it the capacity to constitute the whole*.

To philosophically speculate, what does all this have to do with reality? Rationally conceived, that insight is meaningless. But if the truth of it may be seen through a kind of indescribable intuition, via a novel modality of consciousness, the insight will be received as revelatory liberation. One will come to know their "self" as a slivered portion, a unit of Being that, at the same time, somehow constitutes the Whole Thing.

The first adumbration of this mystical insight was explicated by the neo-Platonic philosopher Plotinus with his considerations on Beauty. In his essay, *On the Nature of Beauty,* he writes: "And on what has thus been compacted to unity, Beauty enthrones itself, giving itself to the parts as to the sum: when it

---

[*] Chrapkiewicz, R.: *Hologram of a Single Photon.* (2016)

lights on some natural unity, a thing of like parts, then it gives itself to that whole."˙ We will return to this shortly.

The fact that a hologram contains — encoded in two dimensions — all the information (at every point on the surface) to construct a three-dimensional figure inspired and lead directly to a fundamental tenet of string theory. First appearing in black hole physics, the **Holographic Principle** holds that the information required to describe a certain volume of space can be encoded on a lower-dimensional boundary of said volume. This idea immediately strikes one with the possibility that our entire universe may be a hologram whose informational content is stored on its boundary, a boundary we know to exist called the **cosmic horizon**.

Quite possibly, the holographic principle may establish itself without a "boundary" at all. Maybe, all that is necessary to build up a vivid universe is entangled information "stored" in nothing. A lower bound that does not need to look anything like the structures it supports. What would that lower bound even look like? Could our familiar four-dimensional spacetime be a projection of entangled information in the vacuum? Are we circles living in Abbott's flatland projecting ourselves as spheres? These are deep questions.

All we do know, and what is becoming clearer by the day, is that the *interactions of information establish relations that in turn solidifies actuality*. Said differently, qubits of information subsisting in the quantum foam come to relate in certain ways, these define its degrees of freedom and congeals energy into "reality." For instance, without some object standing in relation to another, we are not able to ascertain its motion or orientation. An object standing out from a void is just that - lifeless. Add another and we may witness relative motion although if we lived on one of them, we would not be able to discern which is moving with respect to one another. From the frame of reference of either, it will be seen as though the other is in motion. Finally, it is only by adding a third object that we will be able to see which object is moving in which way. By this, we see that it is the establishment of relations that distills coherency into actuality.

This is why all things, including space and time themselves, may emerge from the entanglement relations of the degrees of freedom that structure the indeterminate quantum foam. In fact, the classical, three-dimensional world ruled over by gravity as cosmologically subject to general relativity may be the "bulk" side of a holographic quantum coin. Its boundary side, of course, being the quantum sea of information carrying interfering waves. What might constitute the universe's "reconstructive waves" such that it obtains actuality is consciousness itself.

˙ Plotinus.: *On the Nature of Beauty.* (After Aristotle)

"There is a supreme experience and supreme intuition by which we go back behind our surface self and find that this becoming, change, succession are only a mode of our being and that there is that in us which is not involved at all in the becoming. Not only can we have the intuition of this that is stable and eternal in us, not only can we have the glimpse of it in experience behind the veil of continually fleeting becomings, but we can draw back into it and live in it entirely, so effecting an entire change in our external life, and in our attitude, and in our action upon the movement of the world. And this stability in which we can so live is precisely that which the pure Reason has already given us, although it can be arrived at without reasoning at all, without knowing previously what it is — it is pure existence, eternal, infinite, indefinable, not affected by the succession of Time, not involved in the extension of Space, beyond form, quantity, quality, — Self only and absolute."[*]

Hidden Harmony

Now, what exactly about the universe is holographic? How can all of the information required for its coherent and total evolution exist — always already — at every point of its global Being? How could any definable part of this totality "contain the whole" such that both are in some deeper sense one and the same Thing?

This is a truth not easily seen through the employment of the rationalizing intellect but instead requires the instantiation of a novel state of mind for this revelation is witnessed as self-evident in an "altered state" of consciousness. If one keeps an open mind, the holographic nature of reality can be grasped by a pure intuition through an *epiphanic mode of givenness*. This non-normal mode of conscious awareness is known colloquially as "epiphany" and it is only when this mode is lived-in-and-through for an extended duration that it reveals the true nature of reality as Divine.

It almost always goes without explicit credit, but the nature of some of the deepest insights is arrived at through a rare, but undeniable knowledge of one's unity with the whole of nature. "The highest good," as Spinoza so eloquently pointed out, "is the knowledge of the union that the mind has with the whole of nature."[+] That understanding is "given" to a particular individual through an epiphanic mode of awareness.

The 'highest good' is the epiphany that washes over one, such that they are overcome by an indescribable bliss, and are liberated momentarily by the

---

[*] Ghose, A.: *Life Divine.* (1919)
[+] Spinoza, B.: *Treatise on the Improvement of the Understanding.* (1656)

knowledge that they are truly One with the universe. In the mode of epiphany, having borne witness to a knowledge won in this way, one will never falter from the truth of it and carry it with them always, even if they never come to experience it again. Indeed, the lack of its recurrence may cause some to feel as if they've been forsaken such that they may begin to doubt but it will be a doubt easily pushed aside by memory.

Regardless, a testament and vivid example of the unshakable certitude won by this mode of givenness was revealed in the life of the Italian poet-philosopher Giordano Bruno. Through epiphany, Bruno came to understand a cosmological truth — that the lofty stars in the night sky were, in fact, other suns, each of which, most probably, contained many worlds, not unlike our own. Although unable to *prove* this for others, as is the case with most intuitive insight, Bruno never strayed from claiming this knowledge. He did so in spite of the fact that the Catholic church planned to burn him at the stake for his heretical "theories." In the year 1600, as the flames licked his feet, Bruno never recanted and was burnt to death.

The certainty begetting such an epiphanic mode of givenness is obtained because within the state itself one *is* or *becomes* indistinguishable from the contents of their consciousness. An enantiodromedal transformation takes place wherein subject and object fuse into Unity. To be sure, the crucial insight that inspired and motivated this work came from just such an experience. An experience where the division between subject and object dissolved and the universe as experienced became a single phenomenon.

In the mode of epiphany, the emotional quale of the revelation, of the knowledge thus acquired is genuinely ineffable and of such profound proportion that there exists a "feeling of such overwhelming bliss that the resources of language have been exhausted again and again in the attempt to describe it... This phenomenon usually comes as a tremendous shock. It is indescribable even by the masters of language; and it is therefore not surprising that semi-educated stutterers wallow in oceans of gush. All the poetic faculties and all the emotional faculties are thrown into a sort of ecstasy by an occurrence which overthrows the mind."

Like many, Crowley here attempts the ineffably impossible, to put into words what cannot be put into words. A sentiment better stated by Laozi:

"The Tao that can be told is not the eternal Tao.
The name that can be named is not the eternal name.
The nameless is the beginning of heaven and earth.
The named is the mother of ten thousand things.

Crowley, A.: *Magick - Book Four Liber ABA*. (1994)

Holography / To Write the Whole

Ever desireless, one can see the mystery.
Ever desiring, one can see the manifestations.
These two spring from the same source but differ in name;
this appears as darkness.
Darkness within darkness.
The gate to all mystery."*

Laozi gets it right, but the nature of the feeling and knowledge won by it was put best by the pessimistic philosopher Arthur Schopenhauer in his masterwork *The World as Will and Representation*. It marks one of my favorite passages in all literature and deserves to be quoted at length:

> "When we lose ourselves in the contemplation of the infinite extent of the world in space and time, reflecting on the millennia past and the millennia to come, – or indeed when the night sky actually brings innumerable worlds before our eyes, so that we become forcibly aware of the immensity of the world, – then we feel ourselves reduced to nothing, feel ourselves as individuals, as living bodies, as transient appearances of the will, like drops in the ocean, fading away, melting away into nothing.
>
> But at the same time, rising up against such a spectre of our own nothingness, against such a slanderous impossibility, is our immediate consciousness that all these worlds really exist only in our representation, only as modifications of the eternal subject of pure cognition, which is what we find ourselves to be as soon as we forget our individuality, and which is the necessary, the conditioning bearer and support of all worlds and all times. The magnitude of the world, which we used to find unsettling, is now settled securely within ourselves: our dependence on it is nullified by its dependence on us.
>
> Yet we do not reflect on all this straight away; instead it appears only as the felt consciousness that we are, in some sense (that only philosophy makes clear), *one with the world*, and thus not brought down, but rather elevated, by its immensity. It is the felt consciousness of what the Upanishads of the Vedas repeatedly express in so many ways, but most exquisitely in that dictum already cited above: 'I am all these creations taken together, and there is no other being besides me.' This is an elevation above one's own individuality, the feeling of the sublime."†

---

* Laozi.: *Tao Te Ching*. (4ᵗʰ Century B.C.)
† Schopenhauer, A.: *The World as Will and Representation*. (1818)

Schopenhauer gets it right and recognizes a novel state of knowing that admonishes the fact consciousness forms an intrinsic part of, if not *the*, fundamental aspect of the very ground of existence. Although I do not agree that it is his, our, or my, consciousness that grounds all of existence, just that consciousness *tout-court,* whatever it may, certainly does. Furthermore, it is just this mystical insight that is necessary for the expression of genuine knowledge. "As long as you still feel the stars as being something 'over you,'" Nietzsche wrote, "you still lack the eye of the man of knowledge."*

The mystical intuition regarding the wholeness of reality and the understanding it imparts as to how parts relate to and embody the wholes to which they belong, and vice versa, is the birthright of every living being. The seed of this sentiment is to be found in nearly every great thinker from around the globe.† The immortalizing liberation of this knowledge is a guiding light for all who bear witness to it. May all attain.‡

For those with no intimacy or acquaintance with this kind of knowing I should like to point out that it is often stated that one need be a mystic to understand a mystic. The knowledge of mystical union shares with it the same indescribability and incommunicability as does our emotional sphere and must therefore be experienced first-hand to properly *know* of its validity. Only by each in their own way knowing anger does both know one and the same thing despite neither being able to put into words what anger even is. Again, to know any emotion one must experience for and as themselves and the same goes for

---

* Nietzsche, F.: *Beyond Good and Evil.* (1886)
† A few quotes in this vein. "All goes to show that the soul in man is . . . the background of our being, in which they lie, — an immensity not possessed and that cannot be possessed. From within or from behind, a light shines through us upon things, and makes us aware that we are nothing, but the light is all. A man is the facade of a temple wherein all wisdom and all good abide. . . . When it breathes through his intellect, it is genius; when it breathes through his will, it is virtue; when it flows through his affection, it is love. . . ." - Ralph Waldo Emerson

"In dream you love some and not others. On waking up you find you are love itself, embracing all....When you are love itself, you are beyond time and numbers. In loving one you love all, in loving all, you love each. One and all are not exclusive." - Nisargadatta

"He who experiences the unity of life sees his own Self in all beings, and all beings in his own Self, and looks on everything with an impartial eye." - Siddhartha Gautama

‡ Referring to this transcendental knowledge, "may all attain" was the rallying cry of Aleister Crowley. Another of his was "Do what thou wilt shall be the whole of the law."

the "religious" sphere of experiencing as it is subject to its own forms of knowing and evidence.

Regardless of one's familiarity with altered states of consciousness or alternate modes of givenness, if one has an ounce of intellectual integrity, one will readily admit that the mystery of consciousness is, in general, already quasi-mystical. Science is utterly lost when it attempts to explain it. The reason being is that it is so far removed and different in kind from everything that we know about the universe, despite being the very "thing" that provides us with any and all positive knowledge concerning it whatsoever.

Again, concerning knowledge, we are not limited to the first-order domain of immanent sense perception or the second-order domain of judicative, rational cognition. We also know what anger is like. Some of us know misery, others despair, some hope, and thankfully, many of us know love. These emotions are matters of the heart and the knowledge of them cannot be rationally imparted to others through the limiting use of language or any sort of discourse generally. They must be known in their immediacy.

To that end, a parent knows a kind of love and existential anxiety that non-parents do, and quite simply cannot. Nevertheless, it is through empathy and a strong image of one's self that we arrive at the emotional sphere of our being and derive our morals from it. It is that part of our psyche that can suffer injury and we can feel shame for past acts we cannot undo or despair under the weight of a potentially disastrous future. The morals one arrives at — and devotionally attempts to adhere to — are the result of a continuous evolution of their unique in themselves but also timeless spirit.

Consciousness grounds — and is in direct causal contact with — the universe. If it were not, we would simply be passive watchers of the world. Of course, we are not, although we are constrained by our bodily being and causal forces beyond our comprehension, we nevertheless enact our will freely and possess a degree of freedom in our agency. Mentally, through rational cognition, we accelerate the evolution of thought through purposeful intention, and physically, like the golden touch of King Midas, we alter everything we handle. Unlike King Midas, the result is not always gold.

A phenomenological testament to the freedom of our "will" is begotten by noting the fact that it stands in contradistinction to other elements of conscious awareness and can even be overridden by their sheer force. For instance, we may know and feel shame when we are made aware of a prior mistake and by that very light see ourselves as does the other. A feeling that washes over one, where no amount of willpower can stave it off.

Or consider again anger; at times and despite every effort to keep it buried under the surface, rage can rise like a volcanic eruption and take over our — no longer free — will. Adding to this, all emotions are accompanied by physiological changes such that one in the grip of shame may feel physically ill or in anger, as one's blood begins to boil, their temperature actually rises.

As such, is it not obvious that the affective modality of the will is distinct from the sphere of sense perception or the moral-emotional realm of self-givenness? Is it not undeniable, from primary experience as regards that very experience, that the will is wholly free in its expression? Not to sound pseudo-unscientific but despite all the best neuroscientific evidence that claims the opposite and denies the freedom of the will, what matters is not the evidence of science but the self-evident evidence of experience itself. It is the "place" from which we depart and must always return to. A sentiment best summed up by the Merleau-Ponty quote already stated in the section *The Modern Stance of Reflection.*

The fact that 'consciousness' is in touch with, and captures all of this, speaks of its undeniable fundamentality. A fundamentality that cannot be swept away or reduced to other things.

Given this, one cannot equate the so-called hard problem of consciousness with a similar **problem of the image of the external world**. But generalizing them will help us to make some more claims.

Not only do we fail to explain the genesis of visual perception and the rest of the senses, but we also cannot explain the origin of pangs of hunger and thirst, pain, proprioception, touch or tactioception, sexual pleasure, felt guilt, shame, pride, anger, envy, lust, love, and joy. Intellectually, we willfully manipulate thought, are in contact with logic and number, and creatively, we imagine fictions that break the laws of physics.

Consciousness, the capacity to be aware of and therefore in touch with all these myriad forms — which are really ways in which the world is and ways in which we are — is already "mystical." All of that cannot be explained by appeal to a neurocomputational theory of mind as encoding and decoding information.

Information most certainly has much to do with conscious processing. After all, *every* quantum interaction and process signify the registration and computation of information, and the brain certainly performs this task in a variety of ways. Not only physically, in the form of neuron firings and the cataloging action of countless neurotransmitters and their ontic dynamics, but also imaginatively, in the form of pure rationality and its juggling of concepts as epistemic objects. That is, thought manipulates ideas in the mind, each a concept, a bit of information understood as a fact, and by coupling them together, computes information.

Certainly, information as an insubstantial substance has much to do with consciousness and so too is it registered by the quantum foam and its operator-valued fields. They all exist and so *must* somehow relate even though 'consciousness' remains something "above and beyond" yet also "underneath" it all.

On a multiplicity of magnitudes and scales — a web of awareness interpenetrates the whole of reality. One that computes information as itself and utilizes the indeterminate and open-ended nature of quantum fields as a buffer to course correct the efflorescent evolution of the universe in real-time. You, dear reader, and I, each in our own local way, activate, abstract, and are this — underlying and overarching — holographic field of awareness.
Consciousness is literally, and in a very deep sense, everything we have ever known and is also, in some way that is not immediately clear, the very ground floor of the world.
Now, what might the universe-as-hologram have to do with consciousness, and how does the idea help us? To consider it, let's first examine the historical development of the idea.
The notion that the universe held a holographic nature was first intuited by the French philosopher Henri Bergson. In struggling with thinking of perception as the 'creation' of a mental image, a concept cognitive science still clings to, Bergson writes:

> "The whole difficulty of the problem that occupies us comes from the fact that we imagine perception to be a kind of photographic view of things, taken from a fixed point by that special apparatus which is called an organ of perception-a photograph which would then be developed in the brain-matter by some unknown chemical and psychical process of elaboration. *But is it not obvious that the photograph, if photograph there be, is already taken, already developed in the very heart of things and at all the points of space?*"[.]

It is here that we encounter the kernel of the insight regarding the universe as truly holographic. The nature of reality is such that it is always-already here, there, and everywhere, fully developed "at all points in space," completely and totally a 'thing' that we as agents encounter and directly perceive immanently as the "photograph" itself.
We could say that our human-grade consciousness is confined to a sliver of this field and "operates" and actualizes itself as a *portion* of those infinitely many quantum harmonic oscillators that occupy every point in space while simultaneously admonishing that each harbors the whole as each is not actually

---

[.] Bergson, H.: *Matter and Memory*. (1896) Emphasis mine.

a single solitary thing but is instead a conveniently abstracted "piece" of a timeless and everywhere-the-same sea of sentience.

As for the unique and differently defined (that is, the part of the field you live as you), some of the oscillatory springs may be realized as a mass term or momentum quantity that defines a body and thereby establishes the physicality of the universe as we find it in three dimensions. But further to this, in just the same way that real numbers relate to imaginary numbers as a complimentary dimensioned realized by pointing in a novel "direction," some of the other "springs" may be actualized in the orthogonal dimensions of imaginary consciousness, taking on values that beget qualia. The depth of their establishment thus denotes the degree by which the phenomenon is or may be actualized.

For Bergson, I might add, consciousness is not a thing, nor a substance, but is none other than *the movement of time-memory* as he holds that mind without memory is inconceivable. For me, I see this truth as a quantum system's capacity to retain the information generated by its every interaction and retained within its wavefunction. Every moment, every experience, and every interaction is registered by and codified into the wavefunction and remains there for all time.

The problem today is that modern neuroscience holds true what Bergson considers the problem. It takes the *generation of an image* as an actual fact rather than considering the idea that the image "is already taken, already developed in the very heart of things," meaning that the brain doesn't *generate* anything but instead transmutes a nexus of information exchange into an experience of what is actually *already there*.

We know by decoherence that the environment monitors itself such that the universal, electric matter-field does indeed have a relatively stable although incessantly bubbling configuration. On the whole, we can model the idea that this matter-field is multiply realized on a scale of increasing magnitude. At any one point in spacetime, the scalar mass-density harmonic oscillators of the quantum foam are actualized to take on a somewhat definite value. These points differ as one may be relevant to a solitary atom, another to a molecular bond, and yet another that reflects a heavy element of macroscopic organization.

Consciousness does not cast its web over the mass-density springs even though its actuality remains present to it, underlying it. That is because there exists, *at the very same point, at every point,* infinitely many actualizable quantum mechanical harmonic oscillators ready to take on values but that are each fundamentally the same as they express the summation of properties that populate the quantum foam. One of which — possibly the penultimate One — is awareness.

That free agents are quantum systems characterized as unitary wavefunctions is how consciousness casts its knowingness over its local portion of reality like a blanket of snow. Another image might serve us better, imagine and consider the human being and its nervous system as a kind of semi-stationary antenna. A standing wave pattern distorting and fluctuating within an immense, globally transforming holographic whole. The human being — as a nonlocal quantum system — is a coherent "reconstructive" wave that picks out and specifies a portion of the holographic field.[*]

With regards to perception, sensation, and most vividly, vision, the holographic field picture is this; the total universe is a globally transforming reference wave while conscious agents form local-environment-specific, nonlocal, reconstructive waves. Again, and in this context, by 'nonlocal' I mean only spatiotemporally extended.

Just as the interaction of two complementary, self-same generated and carried waves (that is, they are both "made" of the same substance as they move in and are essentially expressions of the same medium) breathe life into a hologram, so too does conscious awareness arise and obtain actuality at the slivered nexus that separates subject (self) and object (universe as local environment). Again, irretractable, it is here, at the very point where they meet that they are easily seen as different expressions of one and the same thing, mainly, a superposition sum that actualizes the mindfield.

To perhaps model this better, consider the standing wave pattern of an atom-captured electron's wavefunction. It is a globally transforming whole that never actually changes. With macroscopic quantum systems — like those embodied as human beings as individual units — their environmental interactions, however slightly, do change them. The wavefunction as a "standing wave pattern" is indicative of a human person and their awareness is not immediately the same as the physical image of a "standing wave pattern" with its troughs, crests, and nodes.

You see, the dynamics of the maths require that we model waves as oscillating sine-like shapes. But this need not be so as we can consider the nonlocal extension of an "oscillating wave" as a nebulous bubble of field influence — a warped well of actualized property — where its maximum amplitude indicates its effective reach.

Furthermore, a standing wave does not represent an actual change of state, only an *oscillatory* movement within some boundary. In this way, a wavefunction need not be conceptualized over time as an ebb and flow that

---

[*] Robbins, S.: *Time and Memory*. (2012)

variably actuates a field's property, but instead as a continuously variable, fuzzy line of always-already established field activity. The periphery of its affective reach is the place where virtual quanta will begin to be exchanged between the two wavefunctions and a force will begin to develop.

For it is not as if a wavefunction's waveform needs to wait for its oscillation to return to a particular location to be felt. It is not as though the crest needs to reach the crest's amplitude as it is — via superposition — always-already there. A quantum is a bubble of influence that always causes effects in and responds to, ripples in adjacent fields.

From the previous two chapters, we have seen that the mindfield can be characterized by three overarching features. First, like a quantum field, it is an oceanic totality of which we all actuate as our very being. Second, as an exponential movement of proliferating information as its objects come to define themselves through their interactive relations. Both epistemic (mind) and ontic (matter) objects reveal themselves in this way — as the computational registration of endlessly flipping qubits. The "substance" that is information bears within it enough open-ended possibility to entertain this as fact. And third, as holographic, whose reality is but another way to express the **mystic's maxim** for everywhere, everywhen, the fundamental nature of Nature remains unchanged and is — in possibly the most important sense — the same "at all the points of space." You and I are One.

QFT and its infinite substratum grants us a novel image of "the world." It is no place, no total 'thing,' but is instead an ever-blossoming, fractaling flower: The Infinite. An effervescent becoming encompassing countless immaterealities whose anchor is the familiar three-dimensional world of matter and form.

I hope to have shown that some version of panpsychism is the correct view with regards consciousness in the universe. Not a panpsychism that sees all things as imbued with awareness but instead by seeing that everything already takes place upon a universal ground that *is itself awareness.* An electron need not actualize the still deeper property of the quantum foam that we live as awareness, nor will a rock even though it is present upon the mindfield that, like all fields, is simultaneously omnipresent and intertwined through all space. The sea of sentience permeates everything, even rocks, but rocks are not sufficiently complex enough to actuate it although it is nevertheless "there."

Each of us, in our own unique way, are expressions of this one and the same field and so are in the deepest and most meaningful sense possible, identical. We are a Unity disguised as multiplicity.

Holography / To Write the Whole

We are all One. We are large and contain multitudes.[*] Each of us is an infinite being situated in a nesting of infinities. Within, we are without limit, endogenous to us is endlessness. Without, we are without limit, for we are the universe unfolding and have come to reflect upon ourselves.

This image of reality is more akin to the Buddhist's notion of void Śūnyatā and of its interconnectedness as revealed by the myth of Indra's net. The Buddhist scholar Francis Hook writes:

> "Far away in the heavenly abode of the great god Indra, there is a wonderful net which has been hung by some cunning artificer in such a manner that it stretches out infinitely in all directions. In accordance with the extravagant tastes of deities, the artificer has hung a single glittering jewel in each "eye" of the net, and since the net itself is infinite in dimension, the jewels are infinite in number. There hang the jewels, glittering "like" stars in the first magnitude, a wonderful sight to behold. If we now arbitrarily select one of these jewels for inspection and look closely at it, we will discover that in its polished surface there are reflected *all* the other jewels in the net, infinite in number. Not only that, but each of the jewels reflected in this one jewel is also reflecting all the other jewels, so that there is an infinite reflecting process occurring."[†]

An infinity nested in an infinity nested is the Infinite. Each of us, every person, every creature, every self-contained, individuatable, and unitary wavefunction is an infinity unto itself.

On a final note, after all has been said and done, looking back on this endeavor I can see better know what it was I set out to achieve. My project was this: to authenticate consciousness as an irreducible, physically real phenomenon and equate it with other equally fundamental omnipresent objects, namely, quantum fields. To achieve this, I hope to have convincingly established the infinite-dimensional scaffolding that underlies QFT (Hilbert space) as physically actual — not merely a mathematical abstraction — and is thus truly representative of the kind of space in which we live. To set up a novel view of the world as not a place, unified totality, or ultimate "thing" but instead as an Infinite ground of indeterminate potentiality that also harbors the capacity to register and record what it itself distills into determinate actuality.

I hope to have — by accentuating the immateriality at the heart of QFT and explicating a realist interpretation with regards to it — accomplished a

---

[*] Whitman, W.: *Song of Myself.* (1855)
[†] Cook, F. H.: *Hua-Yen Buddhism: The Jewel Net of Indra.* (1977)

completely *flat* ontology. A flat ontology grants to everything we are in touch with (and even those things which we are not) existence. Everything exists equally; material substance does not hold existential precedence.

As for the immateriality and determinate indetermination of the quantum foam, I hope to have shown that it possesses another property, namely awareness. Consciousness is real not simply ideal. Both mind and matter are borne of the self-same Substance, there is no true discrepancy between what is generally considered real and ideal. So-called "ideal" objects — epistemic, observer-dependent, immaterially constituted "things" — are just as real as "real" objects — ontic, observer-independent, materially constituted "things." They are each overlapping and complementary portions of a multidimensional quantum foam. Consciousness is capable of being in, and in fact is, in touch with both as it underlies them. The three are of a similar nature, namely, immaterial. Given this, consciousness should be seen as a timeless, intrinsic part of the cosmos, always-already "there" since the dawn of time, but not yet, or necessarily actualized. One might call this quantum-field-theoretical model of consciousness a novel kind of realist panpsychism. One that grants equal existential value to quantum fields *and* consciousness.

At last, we are, each of us, small islands of rolling Being, endlessly exchanging information with a larger portion of ourselves we barely ever realize is us. As we inexorably move through time, we express the movement that is a universe unfolding. To *flow* along with it, in lockstep and in Time, it is best to do as It does, and simply, *create*. Participate in the efflorescent blossom — the endless blooming of Beauty — that is forever coming to know and reflect upon itself. Always and only ever *as itself* — as *You.*

Glossary

Aether – The **gauge** field of **electromagnetism,** Aristotle's fifth element, and Greek god of Light.

Amplitude – The magnitude of a field's maximum displacement from its central value (usually taken to be zero).

Antiquantum – A **fermion**'s charge conjugate twin with identical **mass** but opposite **charge**. That is, opposite **phase shift** transformations.

Asymptotic Freedom – As the energy scale increases and the corresponding length scale decreases, **quarks** exhibit a property that allows them to "feel free" by not needing to respond to the affects of their brethren. Basically, when two entangled quarks depart one another, they will reach a certain distance apart where their exchange of **virtual gluons** will cause them to be pulled back toward one another. As they careen into one another the felt need to do so will drop as the distance does and will thereby become **asymptotically free**.

Background (In)Dependency Problem – That QFT and GR treat spacetime differently forms this formidable problem.

Being – In this text, capitalized **Being** refers to **ontic** reality. What is determinately the case, what is actual as opposed to potential. It is that portion of the **Infinite** that has solidified itself into actuality by interacting with itself. The **Infinite** is an ungraspable somewhat that keeps open countless avenues of indeterminate potentiality while **Being** represents what has been "etched in stone" and is therefore completely determinate.

Born Rule – By squaring the **wavefunction** to remove negative values, we arrive at a probability distribution for a potential interaction location. A place where a **quantum** may **collapse.**

Boson – Along with **fermion**s, they are one of the two broad classes of **quanta**. Their wavefunction's can be symmetric and are indistinguishable from one another. This allows them to cooperate and sum to increase the **field** strength in that location. **Fermions** are distinguishable because their wavefunction's are anti-symmetric and therefore cannot sum.

Bracketing – Placing aside one's many metaphysical assumptions such that they can focus their **intentionality** on the present moment to examine the invariant structures that underlie or produce consciousness.

Charge – A conserved quantity that manifests as a **quantum**'s ability to produce and feel **phase shifts**.

Chirality – That which "twists rotation again." The extra mobius strip degree of freedom twist that allows **fermions** to turn around twice before returning to themselves. See also **helicity**.

Glossary

Color Charge – Quantum chromodynamics' more complex conserved analog of electric charge.
Collapse – When a **quantum**'s **state** changes instantly and **nonlocally** to assume a more compact state. Usually considered to destroy **superpositions**, but even partial collapse (**decoherence**) possesses this capacity to limit the **state**. **Collapse** is no mere limiting of **state** but a more complete kind of transference or transformation of **state**.
Complexity Platform – The word-smith Terrence Mckenna's name for a holon understood as enabler of further complexity. An object that obtains to participate in the structuring of still greater objects. Like a glucose molecule that enters into a relation with a fructose molecule to become sugar. Sugar then becomes a complexity platform in the domain of cooking say.
Confinement – Marks the fact that **color charged** quanta cannot be isolated but will always be found in pairs.
Consciousness – Bare awareness as such. It matters not of what or to which sense that generated it.
Cosmic Horizon – "As far as anyone can see." The cosmic (or cosmological) horizon is a measure of the distance from which one could possibly retrieve information.
Cosmic Late-Comer Problem – The problem of explaining when and how consciousness came to be by restricting its advent to a moment in the universe's history. It is avoided by assuming its timeless existence as a field.
Cosmological Multiverse Hypothesis – Nothing like the **Many-Worlds** interpretation that assumes infinitely many copies of reality exist, the multiverse hypothesis simply assumes that their may be – informationally closed-off and causally inert – universes next door. That beyond our Big Bang horizon there may actually exist unreachable worlds. Also unlike **Many-Worlds**, this is not unreasonable.
Decoherence – A subtle **measurement**-like process that limits the state of a **quantum** and causes *partial* **collapse**. The effect requires another **quantum** to cause the partial collapse which converts the previously **coherent superpositions** into an entangled **mixture** that correlates the two quanta.
Dirac Equation – An evolution of **Schrödinger's equation** that takes relativistic effects into consideration and reveals the existence of **antiquanta**. It explains the creation and annihilation of the **psi field**'s quanta, whether electrons or positrons.
Eigenvalue/vector - In linear algebra, an eigenvalue is a scalar value associated with a square matrix while an eigenvector is a non-zero vector that corresponds to a specific eigenvalue of said square matrix. Geometrically, an eigenvector represents the direction along which the matrix performs only a scalar stretching or compression transformation, and the eigenvalue represents the scaling factor (the amount of stretching or compression) along that direction.

Glossary

For intuition, one can think of 'eigen' as denoting a range of values rather than s single specific one.

Elasticity – The degree of ease with which a **field** is able to propagate change. Likewise, a measure of a fields ability to resist distortion.

Electromagnetism – The theory of electric and magnetic fields.

Energy – Classically defined as the ability to do work or affect change. But here understood in Heisenberg's sense as a kind of substance. A conserved quantity related to time and can be isolated into types: kinetic, potential, chemical etc.

Entangled Hierarchy – A substantial, entangled object that embodies and achieves actuality in many layers or levels of reality that — like a **strange loop** — folds back upon itself to remain itself. An entangled hierarchy achieves this both by being what it is and what it is not, but also mereologically as an emergent phenomenon that is more than the sum of its parts. An entangled hierarchy may or may not be a **holon**.

Entanglement – The process by which two quanta come together, exert forces on one another, and intermingle to such an extent that after they behave as a single quantum and long after their separation remain instantly interconnected.

Epistemic - Relating to the mind i.e., ideal.

Epistemology – The school of thought that takes "knowledge" as its object. A meta-science that makes knowledge claims about knowledge and seeks to articulate the existence of "mental" objects.

Epoché – A reductive, psychological maneuver that allows us to suspend judgment and dispense with all that we take for granted to instead focus solely on what is "given" to us in the present.

Explicate Order – David Bohm's notion for the common world of our spatially and temporally extended experience. Underlying this perceptual order is an **implicate** (or enfolded) **order**, a deeper level of reality out of which even space and time emerge.

Fermion – Along with **boson**'s, the other broad class of quanta. Fermions are solitary quanta whose states cannot sum. They obey **Pauli's exclusion principle**.

Field (Classical) – An incomplete concept. A physical, albeit immaterial, property or function of space. An entity or "thing" permeating the whole universe whose energies are continuous and does not integrate or explain quantum properties like **entanglement** or **collapse**.

Field (Gauge) - A necessary **interaction field** that allows for the **local symmetry** propagation of **phase shifts** that conserves **charge** to culminate in a felt **force**.

Field (Interaction) - See Field (Guage).

Field (Phase) – The proper name for a **gauge field** as they truly arise from *phase* relations.

Glossary

Field (Quantum) – A more detailed description of a **field** but possibly incomplete. A quantum field is a physical, albeit immaterial, "**operator-valued**" characteristic of the **quantum foam** that permeates the universe and whose energies are quantized as waves. Their effervescent bubbling exhibits spontaneous activity and only have a "lowest energy" state not a "no-energy" state making the universe thus conceived an **immaterial plenum**.
Field of Sense – Markus Gabriel's term for a contextual frame that helps us to see the reality and contours of various objects and their types.
Frequency – The number of cycles per unit time regarding any **oscillation**.
Force – Classically, an external influence on an object that changes its **momentum**. The idea of force in **quantum field** terms is replaced by an **interaction** or **gauge field**. Certain **symmetries** necessitate that they exist.
Function – A relationship or expression involving one or more variables.
General Relativity – Einstein's theory of gravity. Modeled as spacetime curvature or metric field. Take your pick.
Hard Problem of Consciousness – Why, and more importantly, how is it that we are in touch with **qualia** – phenomenal properties that obtain only by being observed in relation with a mind. In a word, how does inert, lifeless matter beget mind.
Helicity – The helical aspect of a wave's propagation. Circular polarization. Both **bosons** and **fermions** possess helicity but only fermions possess **chirality**.
Higgs Boson – The quantum of the Higgs Field.
Higgs Field – The spin zero **boson field** that turns the **vacuum** into a cosmic superconductor, serving as a potential for certain **quanta** to exploit in acquiring **inertia**.
Higgs Mechanism – The coupling of the **field** to other's such that it changes the **elasticity** of the **vacuum** as felt by certain **quanta** engendering them with **mass**.
Hilbert Space – An abstract **vector** space that can be off any number of dimensions.
Hologram – An interference pattern of superposed light recorded on a two-dimensional surface that when properly lit projects a three-dimensional image.
Holography – The process of wave interference (re)construction.
Holomovement – Bohm's name for the 'carrier' of the **implicate order** that accentuates the dynamic nature and wholeness of an unbroken and undivided totality.
Holon – A holon is a whole in and of itself but may also be a part of some larger-order object. It is a thing that has identity and integrity while at the same time is a subsystem of some grander system.
Image of the External World – How the world appears to conscious agents in perception. Philosophically known as a **representation**.

Glossary

Immaterial – Absent **inertia**, lacking **mass**, untouchable, intangible, incorporeal.
Indeterminacy Principle – Heisenberg's great contribution illustrates that the more well-defined a variable is, the less so is its conjugate. Places a lower limit on a material **quantum**'s complementary variables (like position and momentum) such that an increase in the determination of one lowers the determination of the other.
Inertia – The property of **matter** by which it continues in its existing state of rest or uniform motion in a straight line unless that state is changed by an external **force**. Resistance to acceleration, **mass**.
Infinite, the – When capitalized, the Infinite refers to a novel image of "the World." Not as a totality or place, but a fractalian One where infinitely many dimensions of Being simultaneously co-exist - *some* of which are encountered and realized by conscious creatures.
Information – Information can be understood in a myriad of ways, as a measure of communication or data in a message. Physically, as a fundamental, either/or, binary unit of computation. Quantum mechanically, a **qubit** exists in a state of either/or at the same time. With respect to consciousness, this fundamental unit may appear as a fact.
Intentionality – The structure and "directionality" of consciousness. In the subject/object paradigm of **relative consciousness** it is open-ended and inviting. It can be understood geometrically as "horizontal," a two-way street moving bidirectionally from within to without. Or "vertical" when characterized by **transcendental consciousness**.
Intentionality (Horizontal) – The operant mode of everyday, subject/object experience.
Intentionality (Vertical) – The operant mode of religious or spiritual experience that contains its own sphere of evidence. It is characterized morally and can be seen in moments where we strive to "stand tall," "look up to," and keep our head "held high."
Invariant – A **function**, quantity, or **property** which remains unchanged when a specified transformation is applied.
Gauge Field – See Field (Guage).
General Relativity – Einstein's evolution of **special relativity** to include gravity. Can be framed as the curvature of spacetime or a **metric field**.
Gluons – Elementary quanta that "glue" **quarks** together to form nucleons.
Gravity – The curvature of **spacetime** as caused by condensed aggregates of energy-momentum. A warping of the **metric field**.
Implicate Order – David Bohm's notion that there exists an order that is contained or enfolded in every region of space and time. This order possesses a holographic nature in that any part of it is capable of specifying the

Glossary

wholeness and totality of reality. This order is pure form and structure that gives rise to space and time.

Intentional Structure of Representing (ISR) - To be contrasted with the Representational Structure of Intentionality, the ISR denotes the in-to-out directionality of Relative Consciousness.

Incompleteness Theorem - The

Invariance - A quantity or property that remains unchanged after a transformation.

Local State Solution - Art Hobson's name for a proposed solution to **the problem of definite outcomes**. It accentuates the nonlocal character of the entangled measurement state and argues that what is superposed is neither subsystem nor their composition but rather the *correlations between* its subsystems.

Locality - The notion that only objects in direct physical contact can influence one another. This is violated by **entanglement.**

Lorentz Transformation - In **special relativity**, the transformation from one frame of reference to another. Through the transformation one may observe **time dilation, length contraction** and/or a **simultaneity shift.**

Mass - A measure of inertia or resistance to acceleration. A physical quantity that **special relativity** equates with energy.

Magnitude - Expresses the relation of various strengths of certain physical parameters.

Measurement - Any interaction that **collapses** or limits the state of a **quantum**.

Measurement Problem - The issue of trying to make sense of the **measurement state**. It can be subdivided into two related problems, the **Problem of Definite Outcomes,** and the **Problem of Irreversibility.**

Measurement State - An *entangled* superposition state of two or more quanta.

Medium - A substance through which a **wave** can propagate.

Metric Field - The "mother of" or "underlying all" field of gravitation and spacetime itself. It can be understood geometrodynamically as matter/energy's capacity to "curve" the fabric of spacetime.

Minkowski spacetime - The spacetime structure revealed by and underlying **special relativity.**

Mixture - When a **quantum** is in a definite yet unrealized state. In the double-slit with a "which-path" detector, the **entanglement** between them **quantum** and detector **decoheres** (and thereby limits the state) such that the **quantum** goes through one or the other slit — not both. But until it **collapses** into the final screen one cannot tell which slit it came through.

Mode of Givenness - A "way" in which objects, or things of any sort, are "given" within conscious experience.

Glossary

Modern Stance of Reflection – All knowable reality is known first and foremost by and within the confines of the mind. We ponder and posit existence from "within the confines of our heads," so to speak.

Momentum (Angular) – The rotational analog of **linear momentum**.

Momentum (Linear) – A **vector** quantity representing a physical system's "quantity of motion;" the product of **mass** and **velocity**.

Morality of the Mystic – With the knowledge of unity possessed by the mystic, one treats the whole of the world as an extension of themselves and so behaves in accordance with that knowledge.

Mystic's Maxim – Knowing that the true Nature of the Self is the very same I that is Me and You and is God.

Natural Attitude – This psychological gloss paints consciousness and existence as wholly mundane and takes for granted the indefinitely many metaphysical assumptions that underlie "normal" experience. By performing the **epoché,** it may be **bracketed** such that consciousness can turn to analyze itself.

Non-Abelian Phase Symmetry – An "order of operations dependant" **symmetry** of **wavefunction**s that begets **interaction fields** that appear to us as classical **force**s.

Nonlocal – The spatial spread of a unified **wave**. Also, the instantaneous connection shared between **entangled quanta**.

Nonlocality – A violation of **locality**. The direct and instantaneous affect one **quantum** has on another across arbitrary distance. They are only capable of such an effect when they are **entangled**.

Ontic – Relating to the physically and factually *real*. Determinate actuality.

Ontology – The school of thought that takes "existence" as its object of study. It investigates what constitutes **Being**, what counts as an object, and what does it even mean for something "to exist."

Ontological Problem of the Wavefunction – Denotes the scission between the mathematical formalism and its ontological referent and overlap. Basically, is the **wavefunction** a mathematical abstraction or does it represent a physical actuality?

Oscillation – The repetitive motion of a physical system as a function of **frequency**.

Pauli Exclusion Principle – Holds that no two **fermion**s can be in the same state at the same time for if they could their **wavefunction**s would interfere and cancel each other out. At least one **quantum number** must differ between them.

Period – The repetition time of an **oscillation**.

Phase – A point or place on the shape of a wave. Mathematically, an angle that demarcates different stages of an oscillating motion. The overlaying of equal frequency and wavelength phases allows us to see whether they are in step with one another; they will be either "in" phase or "out" of phase. Overlaying phases of differing frequencies and wavelengths allows us to see how they **superpose** to form a novel **waveform**.

Phase Shift – An operation on a **wave**'s **phase** whose change must be communicated via an intermediate **gauge field**. Phase shifts are responsible for **charge**.

Planck's Constant – A universal, physical constant that relates the magnitude of **quantum** effects and brings a discontinuity into physical theory.

Presentational Mode of Givenness – The "way" **ontic** and **epistemic** objects are "given" to conscious agents.

Principle of Equivalence – The "force" of gravity is indistinguishable from acceleration. Physically considered, they are equivalent.

Principle of Relativity – the laws of nature are the same for everyone, no matter which frame of reference they find themselves in. The laws of nature are timeless and ubiquitous.

Problem of Definite Outcomes – A part of the **measurement problem**, also called **Schrödinger's cat**. Solutions to Schrödinger's equation imply an indefinite outcome of superposed states but experiment only ever reveals a single definite outcome. This discrepancy forms the problem. It is solved by the **local state solution.**

Problem of Irreversibility – Schrödinger's equation is reversible and can be shown that it does not increase the entropy of the universe. Decoherence with the environment shows that quanta become so entangled with their surroundings that reversibility becomes impossible.

Problem of the Image of the External World – Best described by a question that relies on neuroscientific dogma, how is it that the brain generates a holographic projection of the world? If that's even the case, is the image inside our skulls or external to it as the image itself? See also the similar notion regarding rationality; the **modern stance of reflection**.

Property – A quality of a physical system that is measurable and whose value describes its state.

Psi Field – The field that harbors electrons and positrons.

Qualia – Individual instances of subjective, conscious experience. The experiential properties of sensations, feelings, perceptions, thoughts, and desires. Qualia are attained by coming into contact with and establishing a relation to consciousness such as the flavor of an apple or its "being" red.

Quanta – The plural of quantum.

Quantize – To assign a discrete value to a physical variable or overlay the structure of quantum mechanics onto classical fields.

Quantum – A unified, specific quantity of space and time extended field energy, characterized as a complex wave.

Quantum Disturbance – An incomplete and unstable **quantum**.

Quantum Field – See Field (Quantum).

Glossary

Quantum Field Theory – QFT is the theory of how quanta behave in fields. It describes the universe as a collection of field distortions waves, and how these behave to bring about the solid world we know. How their symmetry properties dictate what kind of fields there are.

Quantum Foam – The never-still, effervescent lowest energy state of the void/vacuum, of all quantum fields combined.

Quantum Indeterminacy – The fact that coherent quantum systems exist in superposition states of possible outcomes. Prior to measurement, an inherently random process, quanta are indeterminate.

Quantum Jump – A seemingly instantaneous transition from one state to another.

Quantum Mechanics – The state of **quantum** theory before the contributions of Louis De Broglie were solidified by Schrödinger. The "mechanical" precursor theory of the atom that held onto classical concepts like particles but also incorporated quantum properties like "jumps."

Quantum Number - A set of four numbers that are used to describe the state of an electron in an atom. They are referred to as principal, azimuthal, magnetic and spin.

Quantum Potential – A "sub-quantum" field akin to the **Higgs** that harbors an enfolded order out of which manifests the explicate order of experience.

Quantum (Resonant or Real) – A quantum that becomes a field's natural motion. In contradistinction to a **virtual quantum** or disturbance. See also **real quantum**.

Quantum (Virtual) – To be contrasted with resonant or **real quanta**, a virtual quantum is a subtle distortion in the **wavefunction**. Another name for a **disturbance**. A transient or spontaneous fluctuation of a **field** created by a **real quantum**'s **phase shift** with no lasting permanence. Virtual quanta can momentarily break conservation laws and are said to be "off mass shell."

Quarks – Elemental material quanta that are never found in isolation. Quarks, when bound by **gluons**, form nucleons —protons and neutrons — the tempest hearts of atoms.

Qubit – The quantum analog of a classical bit.

Realism – The world, or at least portions of it, exist apart from sentient, knowing agents.

Real Quantum – A field's true **resonance** with a strictly related **mass, momentum,** and **energy.** These are mathematically fixed in a relativistic dispersion relation. **Virtual quanta** do not have to satisfy this relation.

Relative Consciousness – Distinct from **transcendental consciousness**, the "standard" or "mundane" form of everyday consciousness. It is characterized by the subject/object **presentational mode of givenness**. Its intentional structure is "horizontal" it gains its phenomenal weight via the "**natural attitude.**"

Glossary

Renormalization - A technique in QFT used to treat infinities arising in calculated quantities by altering values of these quantities to compensate for effects of their **self-interactions**.

Representation – Because our cognitive faculties inevitably affect and alter the information received by the senses, a **representation** codifies the philosophical notion that we only ever achieve – or better, *know* – a re-presentation of **the world** or some portion of it.

Representational Structure of Intentionality (RSI) – To be contrasted with the **Intentional Structure of Representing**, the RSI denotes the out-to-in directionality of **Relative Consciousness**. It is the **image of the external world** that is "given" to you in perception.

Scalar - A physical quantity possessing only **magnitude**.

Schrödinger's Cat – The name for Schrödinger's original thought experiment that purported to show a deep flaw at the very heart of quantum theory. Resolutions to his equation were thought to predict an impossibility, a zombie cat. Further investigation into this affair reveals that he – and everyone else at the time – misunderstood what his equation was predicting. Which for the time – with almost no understanding of the truly **nonlocal** nature of **entanglement** – is to be forgiven.

Schrödinger's Equation – The heart of quantum theory. A wave equation that describes the time evolution of a non-relativistic, material, quantum system. It captures and describes the stationary states of atoms and molecules in terms of wavefunctions.

Self-Interaction – An elementary **quantum**'s displacement of its own **field**.

Spacetime – The stage and scaffolding of reality. We have many models of it and are not entirely sure which one obtains.

Special Relativity – Einstein's theory of space, time, and constant rectilinear motion. It is defined by the necessity of articulating a preferred frame of reference and promotes the invariant speed of light to a law of nature.

Spin – An **internal degree of freedom** and intrinsic **angular momentum** of some **quanta** best intuited as a **wave**'s helical aspect.

Spinor – A physical quantity requiring a matrix to define. A spinor transforms into its negative if turned 360 degrees but will come back into itself after 720.

Superluminal – Faster than light.

Superposition – The interference pattern created by the process of waves interacting in the same medium.

Superposition Principle – If a **quantum** can be in any one of a multiplicity of **states**, it can be all of them at the same time.

State – The exact condition of a given physical system at a certain time.

State Vector – A vector in a possibly infinite dimensional **Hilbert space** that is used to describe a **quantum state**.

Glossary

Strange Loop – A cyclical, seemingly hierarchical structure that evolves only to always return to itself.
Symmetry – Change Without Change. The notion that an object can remain unchanged after undergoing a change. An **invariant property** is said to be just that (invariant) if a transformation done to some object returns it to itself.
Symmetry (Local) – A **symmetry** whose actualization must be propagated away at the speed of light. Think of rotating a single hexagonal cell in a beehive 60 degrees. On all six sides the adjacent cells will feel this change and register it by themselves turning in place. In turn, this will cause there neighboring cells to register the change and so on. Finally, after the transformation has permeated the whole hive, it will be "returned to itself" and one will not be able to tell if a change has taken place.
Symmetry (Global) – Think of moving the whole universe two inches to the left (whatever that may mean) and you have a global symmetry. A transformation has taken place but nothing has changed.
Time Dilation – The slowing of clocks when the effects of motion and/or **gravity** is considered.
Transcendental Consciousness – A rare, altered state of consciousness characterized by a lived-knowledge as to the wholeness of reality, of the primal fundamental nature of consciousness itself. Lived through in the mode of liberating epiphany, the quality of this state is nothing less than Absolute Bliss.
Velocity – The distance moved by an object over unit time.
Vector – A physical quantity possessing both **magnitude** and direction.
Vector Space – A space consisting of **vectors**, together with the associative and commutative operation of addition of vectors, and the associative and distributive operation of multiplication of vectors by **scalars**.
Vacuum (Sometimes Void/Vacuum) – The carrier of fields. It is devoid of resonant **quanta** but by seething with virtual activity it is better conceived as the **quantum foam**.
Wave – A dynamic shape of a **medium**.
Wave Equation – An equation that describes the interdependence of **frequency** and **wavelength** and how time evolution relates to spatial pattern.
Wavefunction – A **complex** function that forms the primary description of a **quantum's** state.
Wavefunction Collapse – See collapse.
Waveform – The overall shape of a wave.
Wavelength – The **period** of a wave; the distance from one crest to another.
World, the – The hypothesized totality of All-that-Is, complete Reality, the ultimate whole. See my forthcoming article "The Infinite" to see how the concept of which is plagued by many philosophical inconsistencies.

Bibliography & References

Albert, D. Z.: *Quantum Mechanics and Experience.* (1992)
Aristotle.: *Physics.* (Before Jesus/old-as-fuck)
Auyang, S.: *How is Quantum Field Theory Possible?* Oxford University Press. (1995)
Bergson, H.: *Matter and Memory.* (1896)
Bitbol, M.: *Is Consciousness Primary?* (2008)
Bohm, David.: *The Special Theory of Relativity.* (1965)
Bohm, David.: *Wholeness and the Implicate Order.* Routledge. (1980)
Brihad-Aranyaka Upanishad, III.VIII. Translated by Swami Nikhilananda. (1990)
Brooks, R.: *Fields of Color.* (2016)
Carroll, S.: *Something Deeply Hidden.* (2019)
Chrapkiewicz, R.: *Hologram of a Single Photon.* (2016)
Crowley, A.: *Magick - Book Four Liber ABA.* (1994)
Davies, P.: *The Demon in the Machine.* (2019)
Davies, P., Gribbin, J.: *The Matter Myth:* Simon and Schuster. (2007)
de Vries, Jan.: *Old Norse Etymological Dictionary.* Leiden: Brill. (1977)
Durant, Will.: *The Story of Philosophy.* (1926)
Eddington, A.: *The Nature of the Physical World.* (1928)
Einstein, A., Infeld, L.: *The Evolution of Physics.* Cambridge University Press. (1938)
Feynman, R. P.: *The Character of Physical Law.* Cambridge (1967)
Gabriel, M.: *Fields of Sense.* Edinburgh University Press. (2015)
Gabriel, M.: *The Meaning of Thought.* (2020)
Gabriel, M.: *Mythology, Madness, and Laughter.* (2009)
Gabriel, M.: *The World does not Exist.* (2013)
Ghose, A.: *Life Divine.* (1919)
Glattfelder, James, B.: *Information—Consciousness—Reality.* Springer. (2019)
Hobson, A.: *Tales of the Quantum.* Oxford. (2017)
Hobson, A.: *There are no Particles, there are only Fields.* (2012)
Hofstadter, D.: *I am a Strange Loop.* (2007)
Hu, Huping., Wu, Maoxin.: *Spin as Primordial Self-Referential Process Driving Quantum Mechanics, Spacetime Dynamics, and Consciousness.* (2003)
Laozi.: *Tao Te Ching.* (4th Century B.C.)
Lloyd, S.: *Programming the Universe.* (2006)
Maxwell, J, C.: *A Dynamical Theory of the Electromagnetic Field.* (1864)
Maxwell, J, C.: Letter to Frederic Farrar. (1854)
Meillassoux, Q.: *After Finitude.* A&C Black. (2006)
Merrell-Wolff, F.: *Experience and Philosophy.* (1994)
Merrell-Wolff, F.: *Transformations in Consciousness.* (1995)
Mills, R.: *Space Time and Quanta.* (1994)
Nietzsche, F.: *Beyond Good and Evil.* (1886)
Penrose, R.: Mandelbrot ()
Plotinus.: *The Enneads - On the Nature of Beauty.* Penguin. (1991)
Prigogine, I. Stengers, I.: *Order Out of Chaos.* (1984)
Robbins, S.: *Time and Memory.* (2012)

Bibliography & References

Russell, Bertrand.: *The Analysis of Mind.* (1921)
Schlosshauer, M.: *Decoherence and the Quantum-to-Classical Transition.* Springer. (2007)
Schmitz, W.: *Particles, Fields and Forces.* Springer. (2019)
Schopenhauer, A.: *The World as Will and Representation.* (1818)
Schrödinger, E.: *Mind and Matter.* Cambridge. (1958).
Schrödinger, E.: Nobel Lecture. (1933)
Searle, J, R.: *The Rediscovery of Mind.* MIT Press. (1992)
Sebens, C. T.: *How Electrons Spin.* (2018).
Spinoza, B.: *Treatise on the Improvement of the Understanding.* (1656)
Steinbock, A.: *Knowing by Heart.* (2021)
Steinbock, A.: *Moral Emotions.* (2014)
Steinbock, A.: *Phenomenology & Mysticism.* (2007)
Tolle, E.: *The Power of Now.* (1997)
Vedral, V.: *Decoding Reality.* Oxford. (2010)
Voltaire.: *Le Siècle de Louis XIV.* (1751)
Wallace, P. R.: *Paradox Lost: Images of the Quantum.* (1996)
Weinberg, S.: *The Search for Unity.* MIT Press (1977)
Whitman, W.: *Song of Myself.* (1855)
Wilczek, F.: *A Beautiful Question.* Penguin Books. (2016)
Wilczek, F.: *The Lightness of Being.* Basic Books. (2009)
Wittgenstein, L.: *Tractatus Logico-Philosophicus.* (1921)
Zee, A.: *Fearful Symmetry.* Princeton (1999)

www.ingramcontent.com/pod-product-compliance
Lightning Source LLC
Chambersburg PA
CBHW072158200426
43209CB00074B/1942/J